Urban Narratives and the Spaces of Rome

This book foregrounds the works of Pier Paolo Pasolini to study the Roman periphery and examine the relevance of Pasolini's vision in the construction of subaltern identity and experience. It analyses contemporary Italian society to understand the problem of social exclusion of marginal communities.

Narrative studies are at the core of contemporary social science research. This book uses narrative analysis to unpack the deeper meaning of Rome's stigmatized periphery, through an interplay of Italian cinema, literature, and social and political climates. It encourages a positive interpretation of the Roman periphery through a reassessment of its characterization as a homogeneous area of marginality drawing in particular from Pasolini's writings and films on Rome. Pasolini's work left a lasting impact on the modern periphery and the narratives of ordinary citizens, an impact also evident in contemporary street art and popular musical production. His revolutionary vision allows us to appreciate the human and aesthetic character of urban life in regions beyond the city center. The respect for subaltern urban communities encouraged by this book can be extended from Rome to other parts of the world. This book presents an interconnection of social theory, geography, poetry, literature, film and the visual arts to study the experience of life in underprivileged communities.

Written in an accessible style, the book offers a reimagining of the Roman periphery which will appeal to readers in the United States, the UK, Italy, France, Spain, Australia; and other readers with a significant interest in Italian and urban studies, and the works of Pasolini.

Gregory Smith is affiliated with various academic institutions in Italy. He is trained as a social anthropologist from Oxford University. His interests include Italian rural communities and marginal peripheral urban environments. He has published on farming communities in central Italy and on public space in Rome.

Routledge Critical Studies in Urbanism and the City

This series offers a forum for cutting-edge and original research that explores different aspects of the city. Titles within this series critically engage with, question and challenge contemporary theory and concepts to extend current debates and pave the way for new critical perspectives on the city. This series explores a range of social, political, economic, cultural and spatial concepts, offering innovative and vibrant contributions, international perspectives and interdisciplinary engagements with the city from across the social sciences and humanities.

Gentrification as a Global Strategy
Neil Smith and Beyond
Edited by Abel Albet and Núria Benach

Gender and Gentrification
Winifred Curran

Socially Engaged Art and the Neoliberal City
Cecilie Sachs Olsen

Peri-Urban China
Land Use, Growth, and Integrated Urban–Rural Development
Li Tian and Yan Guo

Spatial Complexity in Urban Design Research
Graph Visualization Tools for Communities and their Contexts
Jamie O'Brien

Urban Narratives and the Spaces of Rome
Pier Paolo Pasolini and the City
Gregory Smith

For more information about this series, please visit www.routledge.com/Routledge-Critical-Studies-in-Urbanism-and-the-City/book-series/RSCUC

Urban Narratives and the Spaces of Rome

Pier Paolo Pasolini and the City

Gregory Smith

Routledge
Taylor & Francis Group

LONDON AND NEW YORK

First published 2022
by Routledge
2 Park Square, Milton Park, Abingdon, Oxon OX14 4RN

and by Routledge
605 Third Avenue, New York, NY 10158

Routledge is an imprint of the Taylor & Francis Group, an informa business

British Library Cataloguing-in-Publication Data
A catalogue record for this book is available from the British Library

Library of Congress Cataloging-in-Publication Data
A catalog record has been requested for this book

ISBN: 978-0-367-89319-4 (hbk)
ISBN: 978-1-032-03933-6 (pbk)
ISBN: 978-1-003-01850-6 (ebk)

Typeset in Times New Roman
by KnowledgeWorks Global Ltd.

Contents

Figures

1 Introduction

Narrating the city

Narration is the essence of life. The instruments which make narrative possible are rooted in a geographically situated world. A rich spatial lexicon provides the foundation for rich narrative, which in turn supports strong personal identity. Such identity is expressed through the characterization of the self, enriched by confrontation with an often radically different other. This is the essence of an urban environment. Our consumer world segments experience in a way which facilitates receptiveness to consumer stimuli appealing to a false sense of individuality. Giorgio Agamben (2007) urges us to take individual self-realization seriously, to go beyond superficial appearances. We must free ourselves from mass serialization. Richard Sennet (2008) notes that disorder is fundamental in allowing us to achieve a full personal identity. Cities provide the context for constructive disorder where real citizens strive to perfect their human condition. But modern cities represent an increasingly sanitized setting with weakened capacity to stimulate new reflection on the self and the other. Real cities instead revolve around experiences shared in the public arena. Yet these cities have long been under media attack: in modern cities fear, loathing and privatism all go hand in hand (Lofland, 2009:143).

Ingold (2002) provides compelling evidence of how the operation of viewing space and moving through it involves a visual praxis where cultural knowledge and operational activities are inseparable. Narration is a key element in this visual praxis. Space as an encoded medium allows us to externalize a situated understanding of ourselves in a world we can never understand in a detached way. In advanced industrial societies, our spatial vision is invariably naturalized in ways reflecting the society of the spectacle Guy Debord (1994) theorized. Multiple scopic frameworks shape our vision in ways that reflect ideological, technological, ethnic and other biases (Appadurai, 1996). Ed Soja (1996) notes that these forces lack spatial uniformity, being strongest at the socio-political center of urban environments, and weaker at the perimeter. This geographical reality liberates the spatial margins from governing stereotypes, freeing narrative's creative potential.

The dominant vision of the shared spatial world dissimulates its biases and naturalizes contrived claims. The capacity to resist the dominant vision is what this book explores. The setting is Rome viewed through multiple lenses. One lens is furnished by selected writings and films of Pier Paolo Pasolini, expressing a geographically nuanced narrative style sometimes referred to as environmental literature (Brown and Re, 2008). Strong attention to locational minutiae furnishes a means to evade the center's control, challenging dominant stereotypes, and encouraging reflection on how to interpret competing visions of the physically grounded world. The second lens is provided by the narratives of ordinary citizens living in the urban periphery. A final lens is provided by artists who celebrate perimetral cultural autonomy. These various lenses encourage us to question the idea that space is irrelevant. Ed Soja (1996) speaks of a three-dimensional human ontology: the historical, the sociological and the spatial. None exists without the other.

Life only exists in the telling

This book puts into words experiences gleaned in many years of studying Rome, especially the Roman periphery. I have taught an abundance of courses on urban ethnography, guiding students in exercises that document the experience of the city as a physical place (Smith et al., 2014; Smith, 2017). I have also taught a host of courses on the sociology of Rome, using a perspective drawing significantly from Pier Paolo Pasolini. In these years I have strengthened my conviction that difference in the manner of experiencing and narrating the physical world, between an ordinary citizen and a celebrated author and filmmaker, is one of degree not kind. The narrative strategies implemented by an acclaimed artist are more sophisticated than those deployed by the ordinary city dweller. But the structures of experience and narration are the same. The human trajectory unfolds in a physical world the characteristics of which constitute an active ingredient in narrative used to express experience and identity.

Narration is a universal human proclivity used to define the boundaries of shared consciousness (Ochs and Capps, 2009). Something that cannot be narrated has limited existential claim. Language is the privileged narrative vector, and numerous linguistic studies have devoted attention to the construction of space in narrative. One of the most impressive is Modan's (2007) linguistic ethnography showing how community members of an American inner city construct place through spoken discourse, structuring a moral and social geography with powerful political implications. In a similar vein, Scollon and Scollon (2003) furnish the outline for a discipline they term 'geosemiotics.' As the name suggests, the approach is concerned with the way spatial elements are given relevance in linguistic exchanges. Their interests embrace the language used to describe space, and the social dynamics involved in deploying these descriptions. They divide their investigation into various dimensions. Visual semiotics concerns a grammar of representing meaning potential within a scopic frame. Place semiotics pertains to the broader physical context, 'mountains and rivers, oceans and deserts, cities and farms' (Scollon and Scollon, 2003:19). Interactional semiotics describes negotiated interpretations of the objective world. All of these elements establish a sophisticated framework for orienting the exploration of geographically situated narrative.

These compelling linguistic studies describe the physical world as a reservoir of meaning potential given life through human agency. This is a classical orientation within the social sciences; things lack agency, and can only be given sociological life through human activity. An orthodox Marxist standpoint would view any effort to attribute agency to things as being the result of an over-determined relationship between the individual and the things possessed.[1] This overdetermined relationship is a hallmark feature of advanced industrial society. Recent social theory, however, has advocated the necessity to go beyond this conceptualization, and posit a world where things have their own meaning-generating capacity, exactly as they have their own agency. I have in mind contemporary approaches to environmental studies, which reconfigure the conventional relationship between human actors and the physical world. Bruno Latour in particular suggests that a sociology of things can exist on the same level as a sociology of people. The social can denote '*a type of connection* between things that are not themselves social' (Latour, 2007:5). Things and places can be connected among themselves, like they can be connected to people and communities. Seen in this perspective, the material world is not a passive object, but a social actor with independent agency and personality. This new sociology of people and things identifies the agents of social action as *actants*, which can be people, things or even concepts. Each element is an active ingredient in a shifting social configuration. Each is multifaceted, changing, quixotic; interaction among them yields multiple interpretations and outcomes. In this setting, it is impossible to separate society and things, subject and object. One is embedded in the other. In Latourian terms, the society of things is complementary to that of human subjects. Human language has no monopoly over meaning-generating potential; there is also the language of physical reality.

A useful concept to capture this complex and shifting relationship between people, places and things, is the assemblage, a term first used by Deleuze and Guattari (1987) to denote a bringing together of apparently disconnected parts. The assemblage is an ever-changing entity fusing together the most abstract social forces with the intimate depths of human experience. The contemporary relevance of this concept in the context of the city has been explored by Colin McFarlane (2011), who writes of learning assemblages as fluid and mutually-reinforcing forms of knowledge and practices forged by modern reality dwellers. In the life of the city, there are no predetermined links, no fixed process; every momentary understanding is the product of a dynamic conflict-ridden process. The assemblage captures the fluid character of the context in which contemporary narrative must be assessed. It is a living element given a quality of apparent fixity through the interaction of people, concepts and things in a historically and geographically given setting. Assemblages are spatially grounded and strongly constrained by power relations and narrative structures (McFarlane, 2011:27). Territorialization is an important element, as is deterritorialization. This is precisely what we find here: the territory in all of its expressions has altered over long periods of time. In parallel with this transformation are changing patterns of narration with deep spatial roots. In Rome the narrative turning point occurred at the time of Pasolini's writings, when a new voice transfigured a long-standing negative assessment of the periphery. The modern celebration of the periphery emerges from the dynamics of this assemblage.

Citizens give expression to assemblages in their narrative constructions. These constructions include daily exchanges, stories and accounts, street art and musical production. A complex and enigmatic figure like Pier Paolo Pasolini fits well in the narrative context described here. He came from Bologna, but is widely acclaimed as the bard of the Roman periphery. He termed himself a militant communist, but was not a militant in any party. He defined himself as a Marxist, but had a strong idealist inspiration. He claimed to observe the world with scientific impartiality, yet simultaneously professed to be a poet. He enunciated a belief in a world of natural language, yet transformed our material understanding of the world through a combination of penetrating ethnographic documentation and the poetic redefinition of place. Pasolini is celebrated as a realist, but was no crass materialist. He is a supreme narrator and a master of the assemblage. He is a critical figure in our understanding of the Roman periphery, which has a complex life of its own, with an independent capacity to impose meaning on its residents. The periphery is an actant within a composite assemblage whose reception as a meaningful element in human experience has been decisively influenced by a poetic vision which Pasolini (2005) called a form of heretical empiricism.

The sacred as narrative trope

An idea of the civic sacred permeates the writings of Pasolini as it does the narrative of ordinary citizens when discussing the Roman periphery. One dimension of the civic sacred revolves around the intrinsic character of the human condition, also understood as a physical presence. Another dimension, linked to the first, concerns the sacralizing potential of human sociability. A third dimension is the sacrality of place. The idea of the sacred derives from a religious context, but goes beyond this realm in its civic expression.

Pasolini was an atheist with a fascination for religion, and was active in debate which surrounded the Second Vatican Council in the 1960s (Subini, 2007:63). The debate focused on definitions of the sacred. On the one hand was the notion of sacramental purity, whereby liturgical practices define the sacred. On the other was the sheer sacrality of the human condition, understood both as an individual outside of society, and as a person living their intrinsic social condition. The latter was an innovative contribution made by progressive forces in Catholic debate. Before Vatican II the sacred could be narrowly restricted to the confines of religious practice. Afterward, the church recognized humankind's inalienable social character. One of the key documents produced in the Second Vatican Council, *Gaudium et spes,* identifies the sacred quality of the individual while also calling attention to the sacred quality of the human community (Pope Paul VI, 1965:10). Even more important, was John XXIII's *Pacem in terris* (Pope John XXIII, 2014), which sets out the terms of what would later become Liberation Theology, ultimately ushering in collaboration between Catholic and Marxist-inspired political philosophy.

That the sacred can attach equally to the individual or their social condition finds support in the philosophical writings of Giorgio Agamben. Agamben claims that the sacred in the absolute pertains to the *homo sacer,* the pure human living outside the congregation (cf. Vighi, 2003:108). Exclusion is the foundation for this idea of the sacred; a notion defining the sacred quality of social marginality. This is a common trope in citizen narrative generated in Rome, and a permeating feature of Pasolini's writings on Rome's underclass.[2] The sacred pertains to the isolated individual, but is also characteristic of community life 'directly lived' (Agamben, 2007:79). Just as Pasolini's work is inflected with a declared Marxist element, so Agamben notes the relevance of Marxist interpretations when he describes how life in contemporary society is expropriated by a system of media production. This system transforms the world of direct experience into an image, alienating humankind from its spontaneous condition. The sacred has a dual character, poised between exclusion and inclusion, specifically in terms of the role played by a dominant system of commercial signs, and the resistance the idea of the sacred at times inspires.

If the Marxist idea of alienation (Marx, 1992) attributes implicit importance to the spontaneous human community, the formulations of Émile Durkheim are explicit in identifying its sacred character. These ideas have been significantly reappraised in recent years by neo-Durkheimian scholars (Alexander and Smith, 2005). Rosati, for example, explores ideas of the sacred in the contemporary world, arguing that without sacred symbols and practices there can be no social life (Rosati, 2016:13–14). Human communities elevate the individual to a sacred plane through shared practices. From a Durkheimian standpoint, the sacred stands above the individual, and unites a human community through ritual practice (Durkheim, 2008). Latour has also gone back to these ideas, reformulating them in modern terms which stress not the group, but the process of group formation (Latour, 2007:27). For Latour, the group exists as a performative act; when the performance ends, so the group vanishes. Shared memory makes the performance possible, in a process entangling intimate memories and collective ones, dissolving the contrast between individual and group.

For Durkheim, modernity engenders a type of individualism motivated by instrumental rationality, with the attendant loss of a sacred dimension, be it religious or civil (Rosati, 2016:40). Concern regarding this debasement of the human condition is also expressed by Pasolini, when he writes of the triumph of consumer logic over the myth-making power of poetics (Pasolini, 2018). Primordial forces sustaining the sacred were being destroyed in his day by advanced market capitalism, even among the underclass whose very exclusion had given their community its sacred character.

David Kertzer (1980) demonstrates the explanatory power of Durkheimian theory in his study of grassroots Communist Party activities in Bologna in the 1970s. Party membership, the study argues, is not driven by mentalistic concerns, as much as by the capacity of the party to deploy symbols bringing the community together in a set of ritual practices borrowing partly from those implemented by the Catholic Church. We must recall Gramsci's reflections on the hegemonic capacity of the church (Gramsci, 1982:7). After World War II, these reflections provided the foundation for a political strategy mobilizing a counterhegemonic force in the passive revolution advocated by the Italian Communist Party (PCI). Not by chance, Kertzer puts Durkheim and Gramsci side by side in the opening pages of his study (Kertzer, 1980:vii). Durkheim stresses the importance of periodically reaffirming collective sentiments to ensure the unity of the group. Gramsci echoes this concern when describing a lay religion which sanctifies through faith the unity between a world view and a conforming norm of behavior. Gramsci writes that rather than call this a lay religion, we should speak of ideology or politics.[3]

Following this train of thought, we must note the special status held by the 'community' in western political theory. The notion traces back to our Hellenistic roots, and the paradigmatic contrast between the public and the private. This is a 'great dichotomy,' one that 'subsumes other distinctions and makes them secondary' (Bobbio, 2006:2). The public is endowed with an inherent superior quality, associated with politics and equality: the ideal community. The private is associated with economics and a principle of hierarchy, a concept opposed to the former. The preeminence of the public, and ideal equality, are enshrined in the constitution of the Italian Republic, ideals to which all citizens are exposed. This preeminence is the foundation for a civic sacred which celebrates the community 'directly lived.' The city itself, in our western tradition, is an expression of this ideal, found in such well-established institutions as those associated with the town square, so important in Italy's celebrated urban tradition. Canniffe notes that the piazza is a central feature of any urban theory, representing a space and a value which have survived from ancient times to the modern day, even though often as a memory rather than as a direct experience (Canniffe, 2016:15).

The long tradition of conceptualizing and building cities around preeminent public spaces – a veritable motor for uplifting and transforming society – carries on into the present day, where public spaces are defined as 'primary urban elements' (Rossi, 1982:86). These public spaces may be purposely designed, or simply appropriated, as we see in Pasolini's portrayal of peripheral shanties in Rome's sprawling territory. In either case, they are privileged areas of public life designated for the celebration of the community in its pure sense. Some of the citizen narratives described here offer a definition of the public which draws from the exact wording of the Italian constitution. Italian institutions are engaged in many ways to support the preeminence of the public sphere, as are many citizen groups, and much independent citizen activity. But a gradual erosion of these ideals is nonetheless evident.

A fundamental factor in the Durkheimian idea of the sacred is the capacity of rituals to bind the individual to the group. Ritual sacralizes ordinary human actions by expressing shared values and sentiments. Such rituals are characteristic of traditional face-to-face communities, the kind of society that Pasolini describes in his Roman novels. The persistence of this idea of community in the new millennium is implicitly referenced in the citizen narratives described in these pages. Del Negro (2004) has written a valuable ethnography of Italy's most basic ritual: the stroll through a thus celebrated public space. She describes life in a small town of central Italy in the 1990s, where this public appearance is called the struscio. The sacred quality of the act is evident in the strict conditions which inform how one walks, when, where and with whom. It is also evident in the beliefs associated with the performance; the conviction that this practice elevates the individual to the plane of civility. Even in the poor marginal communities described by Pasolini, there is a clear identification of the public as a sphere of action associated with an implicit idea of respectability, and a distinctive aspiration of equality. Still today this ritual performance defines local identity, and establishes membership of a sacred congregation.

My investigation spans a considerable time period. Pasolini provides a window on life in the underprivileged areas of Rome in the 1950s. Citizen narratives were instead collected in the early years of the new millennium. The context has changed significantly, especially with the declining centrality of the public sphere. Yet the classical ideal of the preeminence of the public over the private is remarkably resilient, celebrated in political discourse, in myriad art forms, in selected cultural practices and in the narratives described here. A recent study of the modern periphery of Rome notes the undeniable erosion of the public domain, but goes on to offer this encouraging observation.

> Yet in the periphery minor public spaces [still] narrate minor public activities which nevertheless persist. Maybe the medieval square and the twentieth-century political identities are lost forever; but for now

street life is alive and supports the multifaceted social interactions
which are at the core of urban sociality.

(Cremaschi, 2013:350)

The persisting centrality of the public is part of a peculiar western assemblage, and, like any assemblage, is as unstable as it is difficult to dissect.
My contention is that places, practices and values are fused together in a
dynamic relationship which revolves around an abiding if varied notion of
the sacred expressed in a distinctively urban narrative setting. Cities are
places of memory, as well as a didactic instrument reminding citizens of
the values which subtend the city's creation (Rowe and Koetter, 1983:121).
Physical place condenses meaning, renewed through shared practices
arising from the objective and perceived character of spatial location. The
supreme expression of public space is the medieval square; but even when
such a space is lacking, the celebration of the public may nonetheless be
present. The celebration of these ideals gives life to a performance which is
the city itself. Echoing Latour's idea concerning the performance of society,
I assert that the city ceases to exist when the performance of the city stops.
Fortunately, evidence marshalled here shows that the city as performance
continues into the new millennium.

Ethnography and positionality

Ethnography aspires to capture in writing the image of a culture. A modern ethnography which is pertinent to our interest in underprivileged urban
environments is Elijah Anderson's (2000) celebrated study of violence in an
American inner city. This refined community study opens with the following statement: 'the aim of the ethnographer's work is that it be as objective
as possible' (Anderson, 2000:11). This positivistic aspiration suggests a contrast with the narrative production of an ordinary citizen, or an artist. Yet
some authors question from a scholarly standpoint the very possibility of
'objective' research. Norman Denzin (2010) raises this question, and challenges the idea of a detached view of society generated by an ostensibly
neutral observer. All human action expresses position, he claims, although,
at times, social science research fails to enunciate its inevitable bias. Denzin
believes that qualitative research, such as ethnography, can only be 'authentically adequate' if it expresses position, and represents the multiple voices
that constitute any social realm (Denzin, 2010:26). Such authentically adequate research should also enhance moral discernment, and promote social
transformation. Pasolini does all of these things.

Pier Paolo Pasolini was first and foremost a poet, a civil poet in one post-
humous characterization (Moravia, 1975). The biases of his highly personal
position as a declared Marxist and homosexual were broadly disclosed.
Pasolini cultivated a fascination for marginal communities, a passion discovered in his teens in the region of Friuli, his mother's birthplace. His

earliest publications concern the people and the dialect of this region. Later, he transferred this fascination for peasant life to the Roman underclass he celebrated in poems, novels and films. His passion for detail was matched by a compelling belief that his artistic production could transform the world. He considered poetry to be a weapon he could use to combat injustice (Pasolini, 2012:196). His work expresses special concern for linguistic usage, and both of his novels on the Roman periphery include appendices translating Roman dialect into the standard Italian. His linguistic documentation was assisted by the citizens he worked with, some of whom, as in Friuli, he encouraged to become artists in their own right.

Pasolini's ethnographic writings from his early years in Rome are contained chiefly in two volumes, *Street kids* (*Ragazzi di vita*) published in 1955, and *A violent life* (*Una vita violenta*) published in 1959.[4] Both are replete with obsessive detail, as well as penetrating use of the local linguistic idiom. The rich documentation of locational circumstances conveys the sacred inflection of the places themselves, representing often desperate settings ennobled by the actions and perceptions of common citizens. Language in its vernacular usage represents a characteristic feature of his writing, expressed in crude simplicity to unveil a truth which exalts bare humanity.

Pasolini had almost finished his studies at the University of Bologna when he was drafted into the fascist army, and then deserted in September of 1943 (Siciliano, 1978:74). It was then that he fled to Friuli, living there as a schoolteacher and Italian Communist Party organizer until 1949. His early literary activities, and first publications, were devoted to celebrating Friulian dialect. Then, in the summer of 1949, an incident involving masturbation with his male school children changed his life. In a remarkable statement filed at the local police station, he candidly professed that the activity had literary aspirations, putting into practice ideas described by André Gide in *The immoralist* (Tonelli, 2015:66). However lofty an aspiration, this activity was completely unacceptable in Catholic Italy. And possibly more damning than the activity itself, was its frank public disclosure.[5] The scandal led to his removal from his teaching post, terminating his only source of economic support. In a letter written at the time, he claims that the real reason for his dismissal was the stir created by the Christian Democrats (DC), which carried all the way up to the Italian parliament (Pasolini, 1986:368). The Christian Democrats used the episode as a way to attack the Italian Communist Party (PCI). Yet, as Tonelli has shown, the morality of the PCI was hardly more open than that of the Catholic Church (Tonelli, 2015:4). Indeed, the shattering surprise for Pasolini was expulsion from the party, ostensibly for behavior that was morally unworthy of a communist. Ellero (2004) has shown convincingly that the real reason for his removal from the party was friction with the regional leadership concerning the autonomous status of the Friulian region. Pasolini was an early advocate of autonomy, owing especially to his interest in the distinctiveness of Friulian language. There was also the question of his brother's death, a resistance fighter in a

Catholic unit who was essentially murdered in 1945 by resistance fighters aligned with the PCI (Siciliano, 1978:92). Pasolini published calls in the local press for a clarification of the episode, and this irritated party functionaries (Ellero, 2004:15).

Pasolini, in a letter addressed to a regional party leader, states that notwithstanding his expulsion he would continue to be a communist in the true sense of the word (Pasolini, 1986:368). Thus began his life as a communist outside the party. These events catapulted Pasolini into the national limelight, laying the foundation for a controversial public reputation that was fully consolidated by the late 1950s.

As a consequence of the masturbation episode, Pasolini was forced to flee, and sought refuge in Rome where his maternal uncle, himself involved in a homosexual relationship, provided a temporary haven. Letters amply document the difficult adaptation to Rome, although already in February 1950 he had begun to adjust, stating that Rome was a place where he could express himself with absolute sincerity (Pasolini, 1986:390). His first home in Rome was in the Jewish ghetto, in the heart of historic Rome (Siciliano, 1978:165). This area is featured in episodes of *A violent life*. Then, in 1951, he secured a job as a schoolteacher in a small town south of Rome, and was able to afford to rent a modest flat in the eastern Roman periphery. This was Rebibbia, a largely self-built area not far from the historic borgata of Pietralata (Fig. 1.1). It was another turning point.

The borgate were the product of the 1931 fascist master plan, built as satellite communities to accommodate citizens displaced by fascist gutting out of the center, as well as poor migrants arriving mostly from the Italian south. While the 12 borgate included in the 1931 plan are borgate strictly speaking, the term as currently used is broadly applied to the periphery as a whole with powerful connotations of exclusion. This is the chief setting for Pasolini's literary and cinematic production from the 1950s up to the mid-1960s. In those years, it was the home of an urban underclass which represented for Pasolini a continuum with Friulian peasant society. In these conditions of social marginality, he could find communities still infused with a primeval past, isolated expressions of pure humanity.

> The universe of the peasant (to which urban subproletariat cultures belong [...]) is a transnational universe; one which does not recognize nations. It is the product of an earlier civilization (indeed an accumulation of earlier civilizations [...]
>
> (Pasolini, 2018:52–53)

From his new home in the periphery, Pasolini would take no fewer than four buses to reach the school in Ciampino, and as many to return (Curcio, 2018:19). Anyone familiar with public transportation in the Roman periphery even today can easily imagine the rugged conditions afforded by this service. Pasolini wrote at the time: 'Life is cruel here in Rome: if one is

Figure 1.1 Pasolini's home in Rebibbia in 1951. (Photograph taken in August 2020 ©
Gregory Smith.)

not tough, determined to struggle, one will not survive' (ibid.). In 1953,
Pasolini entered the film industry, and with this new professional condition
was soon able to move to a more central neighborhood (Fig. 1.2). This was
Monteverde, a middle-class district even in that day, in Via Fonteiana, and
a short walk from the Pamphili public housing project dating to the fascist
period. This project is the primary setting for *Street kids,* a novel portraying
life in what is featured as a remote peripheral slum.

These literary – and later filmic – works document life in the periphery,
accompanied by a highly personal commentary. Although he repudiated
the Neorealist film tradition (Rhodes, 2007:60), Pasolini adopted some of
their techniques. Schoonover (2012) describes the way in which Neorealist
filmmakers furnish an implicit critique of the scenes portrayed in their film.
The critique does not require bringing 'the audience to tears and indignation by means of transference, but, on the contrary, it consists in bringing
them to reflect (and *then*, if you will, to stir up emotions and indignation)
upon what they are doing and upon what others are doing' (Schoonover,
2012:xxii). Pasolini deploys this reflective technique in many of his works.
There is no diegetic commentary on events portrayed in the narrative, there
is instead an effort to create 'a site of reflection, a space of critical alterity,' stirring emotions and inviting indignation (ibid.). The position of the
author is not explicitly stated, but the account is so construed as to make the

Figure 1.2 Pasolini's home in Monteverde in 1953. (Photograph taken in August 2020 © Gregory Smith.)

position clear. This is a recurring feature of Pasolini's narrative and poetic constructions.

Some of Pasolini's works, however, infuse his position in a more obvious way. His short documentary *Comizi d'amore* (*Love meetings*) (Pasolini, 1965) is ostensibly a series of empirical interviews laying bare Italian thinking on sexuality. The importance of the film in documenting this little explored area of Italian life in the priggish 1960s was noted by Foucault in a review published in 1977. In fact, the film generates more than sociological truth through a performance in which Pasolini's façade as the impartial documentary filmmaker serves as an ironic commentary on the findings. At one point in the film, he asks a girl in a Milan dance club what she thinks about sexual abnormality, specifying the term 'sexual inversion.' The girl, apparently unaware of Pasolini's publicly professed homosexuality, claims she knows nothing about these 'abnormalities.' When pressed on the topic, she says she would seek medical treatment for her children if they exhibited this sexual orientation. A young man in the same club says he finds homosexuality 'revolting,' and asks if he, Pasolini, shares the judgment. Here are the interviews.

Interview with a young woman in a Milan dance club.[6]
PASOLINI: But listen, some day you will marry and have children, children who might end up like one of these people [i.e., homosexuals].
YOUNG WOMAN: I hope not.
PASOLINI: Let's hope not, I hope with all my heart. But doesn't one have to understand certain problems so one can treat them?

(Pasolini, 2015:110)

Interview with a young man in a Milan dance club.
PASOLINI: Well, what is your attitude toward people who are not normal [i.e., homosexuals]?
YOUNG MAN: Well, I feel … disgust ….
PASOLINI: You feel disgust?
YOUNG MAN: It's the right attitude, isn't it?

(Pasolini, 2015:109)

Sexuality, and especially homosexuality, is a persisting theme in Pasolini's works. His two early novels written in Friuli and only published after his death, involve an explicit exploration of homosexual practices lived with spontaneous innocence (Pasolini, 2000). In his Roman novels, homosexual relations also figure explicitly, alongside brutal portrayals of heterosexuality. Consider the opening scene of *A violent life*, where the protagonist, a young boy from the borgata, seeks a sexual encounter with a male schoolteacher. The latter is portrayed in an image of touching humanity, poor and unsupported. Bearing in mind the conditions in which Pasolini was dismissed from his position as a schoolteacher in Friuli, the scene suggests an autobiographical source with a sympathetic twist. The humanity of this scene stands in stark contrast to the description of a heterosexual encounter furnished a few pages earlier. In this scene, the boys later involved in the incident with the schoolteacher, witness the activities of a pair of sailors who engage the services of two prostitutes. This sexual exchange is described in such unflattering terms as to leave little doubt regarding Pasolini's judgment. The two sailors are consuming the contracted service just inside a little cave on the banks of the Aniene River, a filthy space strewn with rubbish, used condoms and dirty paper.

The two whores and the sailors had stayed just inside the entrance [of the cave] because there was at least six inches of shit inside, and in the dim moonlight that penetrated down there, you could see them

standing, the whores against the slimy wall, the two sailors over them,
like two lizards hit on the back with a stone.

(Pasolini, 2007:19)

While the chief focus of Pasolini's explorations is not sexuality, the
remarkable feature of his work is the ability to connect such intimate
details as these to a broader context. Latour talks about an interpenetration between the person and the surrounding world. It is senseless to speak
of actors as individuals endowed with independent agency. Instead we
must consider the individual as embedded in a wider network: 'the moving
target of a vast array of entities swarming toward it' (Latour, 2007:46).
The continuous interplay between the individual and the broad circumstances in which their existence unfolds is the critical context of the urban
assemblage.

Pasolini captures a deeper meaning in the lives of the urban underclass,
revealing features that respectable middle-class society failed to appreciate.
More conventional studies of the borgata in those days focused exclusively
on social and economic parameters, creating a knowledge foundation that
would allow policy makers to uplift these communities (e.g., Berlinguer and
Della Seta, 1960). Yet this developmental model, according to Pasolini, fails
to appreciate unique dimensions of underprivileged life which should be
recognized and protected.

Pasolini: Public reception

Pasolini is the object of such prolific scholarly attention that even an international journal specializes in the topic.[7] Workshops and laboratories
are organized every summer at the institute entitled to him in Casarsa.[8]
Scholarly interest is matched by global success as an avant-garde filmmaker
and public personality, both in his own day and in contemporary times. His
life and dramatic death were explored by one of Italy's most celebrated filmmakers, Marco Tullio Giordana (Giordana, 1995). To this domestic interest
we must add international commemoration, as evidenced by the American
film director Abel Ferrara's *Pasolini* of 2014.[9]

Textbooks for Italian junior high students describe Pasolini as the 'poet
of marginal citizens,' and this is how he is usually remembered today (e.g.,
Calvani et al., 2019). The online version of Italy's most important encyclopedia dedicates about two thousand words to the celebration of this author.[10]
The entry describes the discovery of his homosexuality and his flight from
Friuli, and details his main literary and cinematic achievements. His violent
death is mentioned in connection with his 'tumultuous activities' in Rome,
and his choice to bear witness to and defend his radical diversity.[11] Pasolini's
work is termed 'a classic' of the Italian 20th century, and his film *Uccellacci
e uccellini* (*Hawks and sparrows*) (Pasolini, 1966) is claimed to be the country's last 'golden legend,' referencing a celebrated medieval religious text

in an accolade which puts Pasolini's poems, novels and films on the plane of Italy's highest literary achievements. The entry concludes with a comparison between Pier Paolo Pasolini and Italo Calvino, another celebrated 20th-century Italian author.

> Postwar Italy thus has two noble emblems of identity: on the one hand, Calvino with the power of reason and utopia, accessible and light-hearted, the Italy of Ariosto and Gallileo; on the other it has Pasolini, the Italy of Jacopone and Belli, Gioacchino da Fiore: rags and apocalypse.

This powerful endorsement is not, however, fully representative of Pasolini's reception either today or in his own times. Pasolini courted controversy, and his reception was mixed to the extreme. The most blistering attack on Pasolini in his day was that of Asor Rosa, Italy's prominent Marxist literary scholar. Asor Rosa's first book, *Scrittori e popolo*, was published in 1965 when Pasolini was in the full flush of his artistic activity. Asor Rosa's hefty volume reviews various portrayals of popular culture in modern Italian literature, ending with a lengthy assessment of Pasolini, whom he dismisses as a reactionary. According to Asor Rosa, Pasolini's writings consist of simplistic categorical statements, expressing the easy emotion and quick temper that appeal to the Italian petty bourgeoisie. His realist portrayals are 'little more than vulgar naturalism' (Asor Rosa, 1979:393). His revolutionary aims are weak, and his novel *Street kids* is the work of a folklorist with no interest in change: 'Pasolini studies the people of the borgata as an entomologist, with his notebook' (1979:414). Pasolini's life and aesthetic project, Asor Rosa concludes, are failures.

Many modern commentators add to this unflattering picture of Pasolini. Zygmunt Baranski terms Pasolini an 'eccentric middle-class iconoclast,' and claims that the carefully fostered image as a radical is a dissimulation of his true intellectual and political orientation (Baranski, 1999:17). His Marxism is that of a dilettante, and his use of Antonio Gramsci – whom Pasolini claims as the source of his political inspiration – is no more than a contrived projection of his own personality. Joseph Francese (1999) asserts that Pasolini, far from being a Marxist, is a carefully concealed idealist who never shook off the influence of Benedetto Croce.

Yet a salient feature of Pasolini's work as it is remembered today is precisely its political commitment. Benedetti (2012) celebrates Pasolini's writing as 'impure literature,' suggesting a contrast with Italo Calvino, whom she reviles as an expression of 'nihilistic disengagement, and postmodern opportunism' (cf. Re, 2014:100). Lucia Re, in reviewing the debate raised by Benedetti's volume, rejects the simplistic antinomy pitting realist commitment against postmodern disengagement. She notes that Calvino's literary production is largely a work of imagination, while Pasolini's writing and filmmaking are corporeal, a kind of 'thinking through the skin' (Re,

2014:105). Indeed, this corporeal quality can be seen as engendering the sense of the sacred so often associated with Pasolini's writings.

Pasolini of course claimed to be a Marxist, not a scholar of Marx.[12] His conversion to Marx is documented in the 1949 poem 'The discovery of Marx.'[13] A more specific statement concerning the character of his Marxism comes from the poem 'Gramsci's ashes' (Pasolini, 1976). This poem dates to 1954, and his discovery of the Roman underclass. The poem's physical setting reveals Pasolini's corporeal approach: we are standing before Gramsci's tomb in Rome's a-Catholic cemetery. In this poem, Pasolini specifies that he is both for and against Gramsci: he is for Gramsci's attention to social injustice, but against his excessive rationalism. Pasolini wished to create a Marxism of the irrational, and implicitly rejects the very idea of the dialectic which is at the heart of any Marxist political philosophy. Clearly his Marxism is highly poetic.[14]

Pasolini was dismissed from the PCI in 1949, as we have seen. He was later brought back into closer association with the party fold in 1960, when invited to collaborate with a weekly publication close to the PCI, *Vie nuove*. Yet his relations with the PCI remained conflictual. Personal accounts show that in the 1970s Pasolini visited peripheral PCI party cells to express electoral support for the PCI (Paris, 2015). More famously, in June of 1975 he made a public declaration of support for the PCI, expressing all of his dissent, but recognizing that no other political force offered an alternative to the existing political system.[15]

In his weekly collaboration with *Vie nuove,* Pasolini begins to formulate a critique of emerging mass consumer society (Tonelli, 2015:118). In his mature reflections of the 1970s, he defines consumer society as a new form of fascism, destined to sweep away the innocence of peasant life (Pasolini, 2018:22). The only way to combat consumer society, he asserts, is through a communist revolution. Yet this is no endorsement of Stalinism. This he notes in an unpublished newspaper article written in 1973, where he discusses the contrast between development and progress. Development is what industrialists want, a way to produce and market superfluous commodities; it is a right-wing concept (2018:175). Progress is instead an ideal social and political notion embraced by workers and progressive intellectuals (2018:176). Yet in the Russian revolution, as soon as the *progressive* forces prevailed, the leadership called for *development*. This paradoxical choice necessarily put peasant society into competition with western industrial societies, forcing Russia to seek ever more advanced forms of development. But development, according to Pasolini, is not a necessary part of a communist society.

Pasolini's Marxism is highly atypical, and perhaps more comprehensible now than it was then. Today, many progressive movements advocate a return to a simpler life, built on a closer relationship with nature, with the community and local roots. This is progress in Pasolini's term. Development understood as the pure pursuit of consumer logic is instead rejected.[16] This post-Marxist position has broad appeal among a growing number of urban and rural activists today.[17]

'Rome ringed by hell'

Cities are complex phenomena, existing as much in the minds of citizens as they do in place. Silverman (1975) devotes an ethnography to the importance Italians ascribe to living in an urban environment. Cities are a quintessential feature of western civilization, and Italy can easily see itself as the lineal heir to this tradition. Silverman's ethnography captures an image of Italy at just the time when it was experiencing an unprecedented surge in urban growth and the demise of traditional rural communities. Such was the impact of this demographic tidal wave that the sociologist Franco Ferrarotti wrote in 1970 that Italy was a 'neo-urban society' (1979:3). A paradox is involved, for citizens in small towns all over Italy already considered themselves to be iconic urbanites, notwithstanding their diminutive demographic status: the preeminent carriers of a civil lifestyle. After the 1950s, a quantitative factor was added to this qualitative sensitivity.

Italian cities have a historically cultivated pride in their urban status. The most obvious spatial setting is the town itself, sharply contrasted with the rural hinterland, the contado. There are then the squares that make up the town, and such landmarks as townhall or the parish church. These features are found in any town, and even in the peripheral reaches of major cities. Indeed, one of the startling revelations in Pasolini's investigation of the extreme Roman periphery is that even there, in the absence of the most rudimentary urban amenities, a civic sense persists. The physical design of central Rome expresses the quintessential western urban ideal, yet few physical traces spill out into the hinterland. The persistence of a civic sense in these remote areas is conveyed in a letter drafted during the summer of 1952, where Pasolini describes his discovery of the borgate. On the one hand, he characterizes Rome as a city ringed by the hell of the borgate. On the other, he exalts the irrationality and the passion of the city, saying: 'Rome [...] is stupendous these days: the bare persisting heat is what it takes to diminish the excess, to lay the city bare and reveal its lofty aspirations' (Pasolini, 1986:490). The 'lofty aspirations' concern the civic sensitivity which emerges even in the most difficult material conditions. The ring of hell preserves an exquisitely western sense of urbanity.

The ring of hell is a powerful metaphor for social and physical marginality. It also expresses a form of essentialization, constructing a convenient spatial backdrop for Pasolini's novels and films. However, neither then nor today was the city characterized by uniform privilege in the center and uniform exclusion at the periphery. The two-city model, for all its political and poetic merits, fails to stand up to scrupulous sociological analysis, as shown by the work of Martinelli (1964), and later confirmed by John Agnew's important publication on Rome (Agnew, 1995). Yet the idea of the two cities in one was broadly accepted, and had powerful impact on local political debate. The two-city model also finds its way into the sociological literature of extreme marginality in European and North American cities, as seen in the noted study by Wacquant where he puts Italian peripheries in the same class as the American ghetto or the French banlieue (Wacquant, 2008:1). The fact that urban marginality might be different in Rome as compared to Paris was noted years ago by the French geographer Anne-Marie Seronde (1957), who already in Pasolini's day argued that Rome was unique among modern European capital cities in that it alone lacked a banlieue. The Roman periphery was a sparse array of scattered communities with no connection among them. Pasolini's work helped unite this diversity into a symbol of exclusion.

In Rome, scarcity and abundance often coexist in close proximity. One of the main settings for Pasolini's first novel, *Street kids*, is a public housing project situated no more than a five-minute walk from the comforts of Pasolini's own middle-class neighborhood (Fig. 1.3). Today this housing project has been partly privatized, and is home to many socially mobile citizens who are thrilled to be part of an edgy peripheral neighborhood not one kilometer from the historic Aurelian walls. The fiction of the two-city model was given sociological respectability in a fundamental publication of 1970, Franco Ferrarotti's *Roma: Da capitale a periferia*. Ferrarotti notes that the periphery contains areas of privilege, but goes on to support the essentializing view

Figure 1.3 Pamphili public housing project in Monteverde. (Photograph taken in August 2020 © Gregory Smith.)

that the borgata broadly understood is tantamount to being a 'total institution' whose function is to establish political control through exclusion and marginalization (1979:xx). Following Pasolini's usage, he employs borgata as a blanket term to define the whole of the Roman periphery. Written in the aftermath of Pasolini's powerful poetical-ethnographic studies, Ferrarotti gives academic respectability to the notion that the borgata is an area of uniform underprivilege. Ferrarotti could not fail to notice the presence of middle-class communities in the periphery, yet he excludes them from his study. He offers a remarkable explanation for this choice, saying that one of the aims of his work is precisely to 'eliminate' the bourgeoisie living on the outskirts of the city both 'politically and historically' (Ferrarotti, 1979:xix). This powerful rhetoric helped galvanize public support for a more inclusionary urban policy directed at the underprivileged periphery.

The Roman elections of 1976 showed the substantial impact the combined forces of poetry and science had had on local public opinion. Of course, it was not awareness of local issues alone that encouraged the shift to the left after thirty years of conservative Christian Democrat (DC) rule. At the national level, the PCI was on the upswing, taking 34% of deputy seats in the national elections of June 1976, up from 27% in the 1972 national elections. At the local level, the 1976 electoral contest brought to power the first progressive town council in Rome since 1907. The political debate directed attention to the borgata, understood as the poorly-planned periphery. Vittorio Vidotto, the most prominent historian of modern Rome, claims that political success was supported by the images generated in Pasolini's novels and films (Vidotto, 2006:335). Vidotto also notes that Pasolini's portrayal did not reflect the real conditions of the city, either in the 1950s, nor certainly in the 1970s when the left swept to power.

Pasolini was the first writer to reveal the periphery in palpable detail to a middle-class public; a periphery endowed with an aura of sacrality. This portrayal was part of his general vision of changing Italian society. As with the peasants of Friuli, he believed that the peripheral underclass was isolated from the corrupting influence of the consumer economy, and continued to express the values of intrinsic humanity. Its isolation allowed the underclass to live with immediacy and wonder, rather than with reason and calculation. He describes in this connection the response of his long-time companion, Ninetto Davoli, to the experience of seeing snow for the first time, in the mountain town of Pescasseroli. Ninetto was a boy from the borgata whose family were recent migrants from Calabria. For Pasolini, he was the quintessential expression of the peasant-derived underclass.

> Looking up makes your head spin. It seems that the whole sky is falling on us, disintegrating into that happy and stormy feast of Apennine snow. You can imagine Ninetto. No sooner has he perceived the never-before-seen event, that disintegration of the sky on his head, not knowing the obstacles of proper upbringing to the manifestation of his

own feelings, he abandons himself to a completely shameless joy. It has two very rapid phases: first a kind of dance, with very precise rhythmic caesuras [...]. The second phase is oral: it consists of an orgiastic-infantile shout of joy that accompanies the high points and caesuras of that rhythm: "He-eh, he-eh, heeeeeeeh."

(Pasolini, 2005:67)

This spontaneous quality of life, with its emotional release unconstrained by a proper upbringing, is what attracted Pasolini to the periphery. It was the manifestation of a pristine human condition set against corrupt middle-class society. Only in the periphery could he find life lived beyond the superficial appearance of individuality. The center was the world of mass serialization, of alienation. True humanity was the privilege of the excluded. Pasolini's work creates a powerful message and a mandate for autonomous cultural life the traces of which we still find in the Roman periphery.

Discerning the sacred

Pasolini's fascination with pure humanity is seen in his novels' graphic descriptions, and in the role played by place and the human body in his films. The opening shot of the film *Accattone,* for instance, dwells on the enigma of the human face (Rhodes, 2007:40). The bare human body figures prominently in his films. The films and novels also immortalize the physical settings in which life unfolds.

Rome is vast, containing physical traces dating to all periods of western civilization. It is home to examples of urban excellence as it is to areas of tragic neglect. The city as a built environment has exercised the interest of generations of historians, one of the most relevant to our interests being Colin Rowe. Rowe was an architectural historian, and while his chief study of urban planning, *The collage city* (1983), reviews examples of urban excellence from around the world, Rome is his real focus. Rowe was a leading exponent of the figure ground school, expressing interest in the way the built environment envelops the urban dweller. Rowe's urban ideal is a cohesive urban design presenting itself to the viewer as a concerted physical environment conveying an impression of harmony. This is what he calls the 'collage city,' in contrast to the 'conflict city.' The creation of such an environment can result either from a central plan or from concerted activity orchestrated among successive architectural interventions. He believed the first model had a totalitarian character, and supported the alternative democratic solution expressed through sequential sensitivity. His ideas have been extremely influential in planning, and represent a powerful tool in assessing Rome's urban achievements (Ellin, 1996:78). Roger Trancik (1986), a student of Colin Rowe, developed the idea of lost space in an effort to preserve the figure ground effect in urban planning. John Agnew's (1995) critique of Rome's urban growth after 1871 echoes the ideal expressed by the figure ground school. David Mayernik (2013) formulates criticism of modern public spaces in Rome, likewise rooted in the thinking of the school. Sympathy for the figure ground model is understandable; it is attractive from both ideological and aesthetic standpoints. The city thus conceived supports 'the utopian illusion of changelessness and finality, [and can] even fuel a reality of change, motion, action and history' (Rowe and Koetter, 1983:149).

The figure ground model is among the most prominent contemporary guides allowing us to capture the aesthetic meaning conveyed by the city as a physical place. Rowe's model was born of studies carried out in the center of Rome, where it works marvelously. Yet, if one visits almost any of the city's peripheral parts, one will be greeted by a riotous variety of shapes and forms, none of which combine into a harmonious whole. Rowe's aesthetic standard for assessing the city is derived from his reading of Picasso's famous collage of 1912 (Rowe and Koetter, 1983:139). In Rowe's assessment, Picasso is weaving disparate elements together into a unitary aesthetic creation. But this is not the only reading either of urban aesthetics or of Picasso.[18] An alternative reading of Picasso, which can provide the foundation for an alternative urban aesthetics, is that of Jay Bernstein who claims that Picasso was not engaged in an effort to achieve a classical standard of visual harmony, but instead pursues 'painterly parataxis to accomplish systematic disordering [...] Juxtaposition rather than logical conceptual subordination' (Bernstein, 2010:226). This form of parataxis breaks the triple spell of beauty established by tradition, idealization and emotional distance. This work represents, in Bernstein's view, a Copernican revolution, pitting 'the voice of sensuous particularity against abstract rationality' (Bernstein, 2010:213).

In parataxis, there is no subordination of the part to the whole. Nan Ellin describes Rowe as a contextualist, as opposed to advocates of the postmodern city who reject context and favor 'messy vitality over obvious unity' (Ellin, 1999:74). Most of contemporary Rome has achieved this postmodern effect unsupported by any conceptual reflection. While Picasso may be a pioneer in the art of messy vitality, a more explicit theory of the parataxic is better associated with the combines Robert Rauschenberg started to produce in the 1960s (Joseph, 2007). The manifesto of Rauschenberg's artistic practice is not by chance entitled *Random order*, and draws from the raw vitality of the city (Joseph, 2007:121). One wonders if Rauschenberg, who lived in Rome shortly before the manifesto's publication, was influenced by the random order of Rome's uncontrolled urban growth. His combines from the early 1960s, like Rome itself, are 'heterogenous and nonhierarchical' using materials 'accumulated into apparently random or paratactic groupings' (Joseph, 2007:138). Other critics have noted the strong democratic inspiration of Rauschenberg's combines, which express 'an egalitarian respect for individual elements of an artwork, combined with a nonhierarchical attention to the mundane' (Leicht, 2012:74).

Parataxis understood in the way described here, as a tool for motivating a sympathetic vision of an uncontrolled urban periphery, can be held to exist in implicit tension with the idea of Junkspace theorized by Rem Koolhaas (Koolhaas, 2002). Koolhaas claims that in the modern deregulated world, the very possibility of harmonious unity is impossible. In these conditions, all the city builder can do is disregard context and devote attention exclusively to architectural design: 'Junkspace thrives on design, but design dies in Junkspace' (Koolhaas, 2002:177). This perspective degrades the value of the modern city, pushing the viewer away from interest in the urban as an aesthetic construct. Parataxis instead affords a new aesthetic paradigm, which leverages the capacity of the viewer to discern the value of space created through the denial of unity. This aesthetic works well in the Roman periphery, but requires that the viewer be committed to discovering beauty in random vitality.

Cities have a life of their own, sedimenting meaning over long periods of growth and decline. The rise of the Roman periphery was certainly influenced by human interventions, but also shaped by forces and trajectories over which the human community had little control. The interface between the ensuing spatial reality and human cognition is part of a constructed vision reflecting many biases, including aesthetic ones. Pasolini's work is in large measure an aesthetic project that wishes to give new meaning to the periphery, redeeming it from conventional repudiation. The success of his project in historical terms can be measured by the aesthetic value that modern dwellers of the periphery attribute to their environment. Cognitive mapping can help us access this perception of value, in a process of which Kevin Lynch was the undisputed master. Starting in the 1950s, Lynch shows how the city 'speaks' to the observer (Lynch, 2002). Naturally, if the image of the city is debased to the point of loathing, and its shared value denied, the language will be offensive and the cognitive map nugatory. Pasolini aimed to overcome the vilification of the periphery, redeemed in a fresh vision of the human condition witnessed beyond the reach of artificial privilege. Citizen-derived topographical narratives described in these pages show that Pasolini's aesthetic project, either directly or indirectly, has been in large measure a success.

Reflection on the role of citizens in giving life to the city has a long tradition. This life-giving activity was the focus of the Italian neo-situationists, whose celebration of the city as a shared space was an artistic project (Bandini, 1977). A similar celebratory practice is today advocated by activists who wish to recapture an unmediated image of the urban environment (Careri, 2006). The practice of engaging the city as a rich and varied physical environment is an everyday pursuit for many ordinary citizens, as well as street artists, street poets, and rap artists. All of these citizens celebrate the city as a performance, today threatened by rising privatism. Pasolini had an understandably pessimistic view of how ordinary citizens can continue to live experience with direct simplicity in a world increasingly dominated by commercial signs. Yet history is fashioned by a process of action and reaction, and the ebb and flow of modern times may be swinging back in favor of the civic culture which so impassioned Pasolini, and still persists today at the heart of the western world.

The structure of the book

Chapter 1 establishes the character of the investigation. Chapter 2 provides background on the city of Rome, exploring its spatial organization. It furnishes details on the rise of the periphery, illustrating the complex character of the city's growth. The topic is vast, and here I focus on features pertinent to the narratives explored in different parts of the book. In particular I try and show how the city-centered model of urban growth misrepresents a torturous historical process. In the modern period, the city of Rome spilled out into a countryside which evolved independently of the city's growth as such. The countryside was the object of contested meaning long before it became the urban periphery, laying the foundation for modern debate on the significance of the urban fringe. Chapter 3 assesses Pasolini's value as an ethnographer of Rome, drawing in particular from his two Roman novels, *Street kids* and *A violent life*. It ends with a consideration of how society and space in the periphery are characterized by a kind of parataxis positively assessed in both connections. It draws attention to the problem of how to reconcile development with progress in the effort to promote positive change without destroying unique features of the periphery. It also makes comparisons with Naples, and American inner cities, to capture elements common to the condition of marginality. Chapter 4 starts with a reflection on poems contained in the 1957 collection *Gramsci's ashes* to understand Pasolini's effort to redefine the semantics of the periphery. By linking this discussion to contemporary street art, we are able to see the continued relevance of these ideas. The rest of the chapter focuses on the film *La ricotta*, exploring Pasolini's thinking about the civic sacred, reviewing in particular connections with Italy's long tradition of visual arts. Chapter 5 presents findings from an exercise carried out on the western periphery of the city. This was in 2013 when I was asked by a local borough to organize a workshop in which I had citizens produce maps and narratives intended for use in a public theater performance. The chapter reflects on the way citizens generate an understanding of the periphery. Chapter 6 focuses on what I term the urban arts, exploring the issue of the autonomy of cultural production in the urban periphery. It devotes special attention to contemporary popular music claiming to draw inspiration from Pasolini. In particular, it reviews the music of Vinicio Capossela as representing a postmodern extension of Pasolini's modernist project. I claim that the modernist view prevails in the contemporary periphery. The chapter also explores rap music produced on the urban fringe, comics and selected examples of street art. Chapter 7 looks at political movements which characterize postwar Rome. It shows that the idea of the community as a rallying point for resistance against market forces has long-standing progressive prominence. The chapter ends with a consideration of Pasolini's relevance for urban environments in other parts of the world, focusing on the contemporary uses of subaltern urbanism.

Acknowledgments

The ideas expressed in this volume emerged over a long period of reflecting, writing and teaching about the Roman periphery. I have incurred many debts in the process. I have had the pleasure of exploring these issues with many students over many years, and I thank them for their enthusiasm and forbearance. My urban studies experiences evolved especially thanks to my association over many years with Cornell in Rome, where I must thank in particular Mildred Warner, Nancy Brooks and William Goldsmith for multiple forms of support. At Temple University in Rome, I have had the opportunity to teach various courses on Italian popular culture, and some of these reflections have found their way into this volume. I must thank in particular Kim Strommen, Hilary Link, and Emilia Zankina. At the University of California Rome Center, I have taught many courses using the works of Pasolini as a window on postwar Italian society, and have benefitted from the support in particular of Julia Hairston and Paolo Alei. Other friends and colleagues who have provided support and advice include Lucia Re and John Agnew of UCLA, Jon Snyder of UC Santa Barbara, Marco Cremaschi of Sciences Po, and Gilda Berruti from the University of Naples Federico II. An important part of the reflection was perfected through the Biennial for Public Space where I had the pleasure of working with Mario Spada and Pietro Garrau. I have also received much support from members of the broader Roman community, including Lucia Cuffaro of the Italian Degrowth Movement, Riccardo dell'Aversano of Brasca Records, Mario d'Amico of the Anonymous Painters of Trullo, and Monica Melani of the Mitreo Theater at Corviale. On a personal front, I have benefitted greatly from conversations with Haydir Majeed about Neorealism in the world outside of Italy. I must thank my wife Claudia Venditti for her patience and support over many years of exploring the Roman periphery. Finally, I have benefitted greatly from two anonymous reviewers at Routledge, and from the professionalism of the commissioning editor, Faye Leerink. Naturally, I take full responsibility for any failings that may be contained in this study.

Notes

1 Maurice Block in his Marxist study of ownership in Madagascar explores the circumstances in which property is 'misrepresented' as a relationship between people and things (Block, 2004:221). The position is antithetical to that proposed here.

2 Throughout I use the term 'underclass' to translate Pasolini's sottoprole-
   tariato, rendered with more technical accuracy by the cumbersome term
   subproletariat.
3 The broad social foundation of the Durkheimian conception of the sacred is
   noted by Bellah, who writes that for Durkheim 'the religious and the social are
   almost interchangeable' (Bellah, 2005:185).
4 Reference made to these volumes in the present work, unless otherwise specified, is to the English-language translations, respectively *Street kids* (2016,
   translated by Ann Goldstein) and *A violent life* (2007, translated by William
   Weaver).
5 Renga (2012) discusses homosexuality in Italy in the 1970s, saying that
   the public environment was so hostile even then that homosexuality
   was 'abjected,' namely, so unthinkable as to be shunned from public
   discourse.
6 All translations from the Italian are by the author unless stated otherwise.
7 *Studi Pasoliniani* is an international peer-reviewed journal (https://www.
   libraweb.net/riviste.php?chiave=26 - Accessed October 5, 2019).
8 An example of these workshops can be found here: http://www.centrostudipi-
   erpaolopasolinicasarsa.it/centro-studi/attivita/incontri/ (Accessed October 5,
   2019).
9 https://www.the-numbers.com/movie/Pasolini-(France)-(2015)#tab=interna-
   tional (Accessed October 5, 2019).
10 http://www.treccani.it/enciclopedia/pier-paolo-pasolini/ (Accessed October 5,
   2019).
11 Major controversy surrounds the circumstances of his death, controversy
   which has received extensive treatment in writing and film. Macciocchi
   claimed at the time that Pasolini was 'assassinated by society in a savage act
   of self-defense' (Macciocchi and Repensek, 1980:11). The self-defense was
   against the hypocrisy and corruption that Pasolini believed permeated Italian society. Macciocchi notes that her bold statement was immediately con-
   demned by the Church, moralists and the Italian Communist Party. Siciliano
   mentions theories according to which his assassination was in fact a staged
   suicide (1978:389).
12 Pasolini's personal library, catalogued after his death, contains mostly works
   of literature (Chiarcossi and Zabagli, 2017). There is not a single volume by
   Antonio Gramsci, and of Marx all we find is the *Communist manifesto* in the
   1947 Italian translation.
13 Published in the 1958 collection of poems entitled *L'Usignolo della Chiesa
   Cattolica*.
14 Tonelli notes that anyone who is looking for a cohesive political position in
   regard to communist orthodoxy will be disappointed by Pasolini (2015:121).
   She also discusses Pasolini's efforts to reconcile religious irrationality with
   Marxist rationality.
15 Published in the PCI newspaper *L'Unità* on June 10, 1975.
16 This is the argument of Serge Latouche's *Farewell to growth* (2015) which
   denounces both neoliberal and Marxist political economies. We will return to
   this in the last chapter.
17 In a rural setting, which also involves Rome, this innovative political position
   finds expression in the concept of social agriculture. For a discussion of some
   current trends see Smith and Berruti (2019).
18 It is only distantly connected to our concerns, but worth noting that Pasolini
   commented on Picasso's works in his poem by the same name of 1953 (Pasolini, 1976). Pasolini claims that Picasso's work lacks the vitality of ordinary
   people which 'explodes happily out into placid festive streets.'

Bibliography

Agamben, Giorgio (2007) *The coming community*. Translated by Michael Hardt. Minneapolis: University of Minnesota Press.

Agnew, John A. (1995) *Rome*. Oxford: John Wiley.

Alexander, Jeffrey C. and Philip Smith (2005) 'Introduction: the new Durkheim,' in Alexander, Jeffrey C. and Philip Smith (eds.) *Cambridge companion to Durkheim*, pp. 1–37. Cambridge, UK: Cambridge University Press.

Anderson, Elijah (2000) *Code of the street: decency, violence, and the moral life of the inner city*. New York: WW Norton & Company.

Appadurai, Arjun (1996) *Modernity at large: cultural dimensions of globalization*. Minneapolis: University of Minnesota Press.

Asor Rosa, Alberto (1979 [1965]) *Scrittori e popolo. Il popolismo nella letteratura italiana contemporanea*. Rome: Savelli.

Bandini, Mirella (1977) *L'estetico, il politico, da Cobra all'Internazionale situazionista (1948-1957)*. Rome: Officina Edizioni.

Baranski, Zygmunt (1999) 'Pasolini, Friuli, Rome: philological and historical notes,' in Baranski, Zygmunt (ed.) *Pasolini old and new: surveys and studies*, pp. 252–280. Dublin: Four Courts Press.

Bellah, Robert N. (2005) 'Durkheim and ritual,' in Alexander, Jeffrey C. and Philip Smith (eds.) *Cambridge companion to Durkheim*, pp. 183–210. Cambridge, UK: Cambridge University Press.

Benedetti, Carla (2012) *Pasolini contro Calvino: per una letteratura impura*. Turin: Bollati Boringhieri.

Berlinguer, Giovanni and Piero Della Seta (1960) *Borgate di Roma*. Rome: Editori Riuniti.

Bernstein, Jay M. (2010) '"The demand for ugliness": Picasso's bodies,' in Bernstein, Jay M. and Thierry de Duve (eds.) *Art and aesthetics after Adorno*, pp. 210–248. Berkeley: The Townsend Center for the Humanities.

Block, Maurice (2004) *Marxist analysis and social anthropology*. London and New York: Routledge.

Bobbio, Norberto (2006) *Democracy and dictatorship. The nature and limits of state power*. Translated by Peter Kennealy. Cambridge, UK: Polity Press.

Brown, Patrick and Anna Re (2008) *Italian environmental literature: an anthology*. New York: Italica Press.

Calvani, Vittoria, Chiara Ferri and Luca Mattei (2019) *Nuovo amico libro. Con letteratura*. Milan: Mondadori Scuola.

Canniffe, Eamonn (2016) *The politics of the piazza: the history and meaning of the Italian square*. London and New York: Routledge.

Careri, Francesco (2006) *Walkscapes. Camminare come pratica estetica*. Turin: Einaudi.

Chiarcossi, Graziella and Franco Zabagli (2017) *La biblioteca di Pier Paolo Pasolini*. Florence: Gabinetto Scientifico Letterario G.P. Vieusseux.

Cremaschi, Marco (2013) 'Contemporary debates on public space in Rome,' in Smith, Gregory and Jan Gadeyne (eds.) *Perspectives on public space in Rome, from antiquity to the present day*, pp. 331–350. Farnham: Ashgate.

Curcio, Valerio (2018) *Il calcio secondo Pasolini*. Correggio: Aliberti.

Debord, Guy (1994 [1967]) *The society of the spectacle*. Translated by Ken Knabb. London: Rebel Press.

Deleuze, Giles and Félix Guattari (1987 [1980]) *A thousand plateaus: capitalism and schizophrenia*. Translated by Brian Massumi. Minneapolis: University of Minnesota Press

Del Negro, Giovanna P. (2004) *The passeggiata and popular culture in an Italian town*. Montreal and Kingston: McGill-Queen's University Press.

Denzin, Norman (2010) *The qualitative manifesto. A call to arms*. Walnut Creek: Left Coast Press.

Durkheim, Émile (2008) *The elementary forms of religious life*. Translated by Joseph Ward Swain. Chelmsford: Courier Corporation.

Ellero, Gianfranco (2004) *Lingua poesia autonomia: il Friuli autonomo di Pier Paolo Pasolini: 1941-1949*. Coderno di Sedegliano: Istitut Ladin-Furlan Pre Checo Placerean.

Ellin, Nan (1996) *Postmodern urbanism*. Princeton: Princeton Architectural Press.

Ferrara, Abel (dir.) (2014) *Pasolini*. Paris: Capricci Films.

Ferrarotti, Franco (1979 [1970]) *Roma: da capitale periferia*. Bari: Laterza.

Foucault, Michel (1977) 'Les matins gris de la tolérance,' *Le Monde*, 23 March.

Francese, Joseph (1999) 'The latent presence of Crocean aesthetics in Pasolini's "Critical marxism",' in Baranski, Zygmunt (ed.) *Pasolini old and new: surveys and studies*, pp. 131–162. Dublin: Four Courts Press.

Giordana, Marco Tullio (dir.) (1995) *Pasolini: un delitto italiano*. Rome: Cecchi Gori Group.

Gramsci, Antonio (1982 [1971]) *Selections from the prison notebooks of Antonio Gramsci*. Edited and translated by Quintin Hoare and Geoffrey Nowell Smith. London: Lawrence and Wishart.

Ingold, Tim (2002) *The perception of the environment: essays on livelihood, dwelling and skill*. London and New York: Routledge.

Joseph, Branden W. (2007) *Random order: Robert Rauschenberg and the neo-avant-garde*. Cambridge, MA: The MIT Press.

Kertzer, David I. (1980) *Comrades and Christians: religion and political struggle in communist Italy*. Cambridge, UK: Cambridge University Press.

Koolhaas, Rem (2002) 'Junkspace,' *October*, Spring, pp. 175–190.

Latouche, Serge (2015) *Farewell to growth*. Translated by David Macey. Cambridge, UK: Polity Press.

Latour, Bruno (2007) *Reassembling the social: an introduction to actor-network-theory*. Oxford: Oxford University Press.

Leicht, Alexander (2012) *The search for a democratic aesthetics: Robert Rauschenberg, Walker Evans, William Carlos Williams*. Heidelberg: Universitätsverlag.

Lofland, Lyn H. (2009) *The public realm. Exploring the city's quintessential social territory*. New Brunswick: Transaction Publishers.

Lynch, Kevin (2002) *City sense and city design. Writings and projects of Kevin Lynch*. Edited by Trebid Banerjee and Michael Southworth. Cambridge, MA: The MIT Press.

Macciocchi, Maria-Antonietta and Thomas Repensek (1980) 'Pasolini: murder of a dissident,' *October*, Vol. 13 (Summer), pp. 11–21.

Martinelli, Franco (1964) *Ricerche sulla struttura sociale della popolazione di Roma (1871-1961)*. Pisa: Libreria Goliardica.

Marx, Karl (1992) *Early writings*. Harmondsworth: Penguin UK.

Mayernik, David (2013) 'Contemporary debates on public space in Rome,' in Smith, Gregory and Jan Gadeyne (eds.) *Perspectives on public space in Rome, from antiquity to the present day*, pp. 301–330. Farnham: Ashgate.

McFarlane, Colin (2011) *Learning the city. Knowledge and translocal assemblage.* Sussex: Wiley-Blackwell.

Moravia, Alberto (1975) 'Ma che cosa aveva in mente?,' *L'Espresso*, 9 November.

Modan, Gabriella Gahlia (2007) *Turf wars. Discourse, diversity and the politics of place.* Oxford: Blackwell Publishing.

Ochs, Elinor and Lisa Capps (2009) *Living narrative: creating lives in everyday storytelling.* Cambridge, MA: Harvard University Press.

Paris, Renzo (2015) *Pasolini: ragazzo a vita.* Rome: Elliot Edizioni.

Pasolini, Pier Paolo (1958) *L'usignolo della Chiesa Cattolica.* Milan: Longanesi.

Pasolini, Pier Paolo (dir.) (1961) *Accattone.* Rome: Cino Del Duca.

Pasolini, Pier Paolo (dir.) (1963) *La ricotta.* Rome: Arco Film.

Pasolini, Pier Paolo (dir.) (1965) *Comizi d'amore.* Rome: Titanus.

Pasolini, Pier Paolo (dir.) (1966) *Uccellacci e uccellini.* Rome: Arco Film.

Pasolini, Pier Paolo (1976 [1957]) *Le ceneri di Gramsci.* Milan: Garzanti.

Pasolini, Pier Paolo (1986) *Lettere 1940-1954 con una cronologia della vita e delle opere.* Edited by Nico Naldini. Bari: Einaudi.

Pasolini, Pier Paolo (2000 [1982]) *Amado mio.* Milan: Garzanti.

Pasolini, Pier Paolo (2005 [1972]) *Heretical empiricism.* Translated by Ben Lawton and Louise Barnett. Cambridge, MA: New Academic Publishing.

Pasolini, Pier Paolo (2007 [1959]) *A violent life.* Translated by William Weaver. New York: Carcanet.

Pasolini, Pier Paolo (2012 [1964]) *Poesia in forma di rosa.* Milan: Garzanti.

Pasolini, Pier Paolo (2015) *Comizi d'amore.* Edited by Graziella Chiarcossi and Maria D'Agostini. Rome: Contrasto Books.

Pasolini, Pier Paolo (2016 [1955]) *Street kids.* Translated by Ann Goldstein. New York: Europa Editions.

Pasolini, Pier Paolo (2018 [1975]) *Scritti corsari.* Milan: Garzanti.

Pope John XXIII (2014 [1963]) *Pacem in terris.* Rome: Libreria Editrice Vaticana.

Pope Paul VI (1965) *Pastoral constitution on the Church in the modern world, gaudium et spes.* Vatican: Documents of the Second Vatican Council.

Re, Lucia (2014) 'Pasolini vs. Calvino, one more time: the debate on the role of intellectuals and postmodernism in Italy today,' *MLN*, 129:1:99–127.

Renga, Dana (2012) 'Oedipal conflicts in Marco Tullio Giordana's "The hundred steps,"' *Annali d'Italianistica* Vol. 30, *Cinema Italiano Contemporaneo*, pp. 197–212.

Rhodes, John David (2007) *Stupendous, miserable city: Pasolini's Rome.* Minneapolis: University of Minnesota Press.

Rosati, Massimo (2016) *Ritual and the sacred: a neo-Durkheimian analysis of politics, religion and the self.* London and New York: Routledge.

Rossi, Aldo (1982) *The architecture of the city.* Cambridge, MA: MIT Press.

Rowe, Colin and Fred Koetter (1983) *Collage city.* Cambridge, MA: MIT Press.

Schoonover, Karl (2012) *The neorealist body in postwar Italian cinema.* Minneapolis: University of Minnesota Press.

Scollon, Ron and Suzie Wong Scollon (2003) *Discourses in place: language in the material world.* London and New York: Routledge.

Sennet, Richard (2008 [1970]) *The uses of disorder: personal identity and city life.* New Haven: Yale University Press.

Seronde, Anne-Marie (1957) 'Rome capitale sans banlieues,' in *Proceedings of IGU Regional Conference in Japan 1957*, pp. 460–465. Tokyo: Organizing Committee of IGU Regional Conference

Siciliano, Enzo (1978) *Vita di Pasolini*. Milan: Rizzoli.

Silverman, Sydel (1975) *Three bells of civilization: the life of an Italian hill town*. New York: Columbia University Press.

Smith, Gregory (2017) 'The pedagogy of an urban studies workshop focused on age-friendliness in selected Rome neighborhoods,' *Urbanistica Tre*, 14:25–32.

Smith, Gregory, Mildred Warner, Carlotta Fioretti and Claudia Meschiari (2014) 'Rome undergraduate planning workshop: a reflexive approach to neighborhood studies,' *Planning Theory and Practice*, 15:1:9–25.

Smith, Gregory and Gilda Berruti (2019) 'Social agriculture, antimafia and beyond: toward a value chain analysis of Italian food,' *Anthropology of Food*, 13 October.

Soja, Edward W. (1996) *Thirdspace. Journeys to Los Angeles and other real-and-imagined places*. Oxford: Blackwell Publishing.

Subini, Tomaso (2007) *La necessità di morire: il cinema di Pier Paolo Pasolini e il sacro*. Rome: Ente dello Spettacolo.

Tonelli, Anna (2015) *Per indegnità morale: il caso Pasolini nell'Italia del buon costume*. Bari: Laterza.

Trancik, Roger (1986) *Finding lost space: theories of urban design*. Hoboken: John Wiley and Sons.

Vidotto, Vittorio (2006) *Roma contemporanea*. Bari: Laterza.

Vighi, Fabio (2003) 'Pasolini and exclusion: Zizek, Agamben and, the modern sub-proletariat theory,' *Theory, Culture and Society*, 20:5:99–121.

Wacquant, Loïc (2008) *Urban outcasts. A comparative sociology of advanced marginality*. Cambridge, UK: Polity Press.

2 The changing faces of the periphery

Concentric rings

Piero Maria Lugli describes ancient Rome as being inscribed in a series of concentric rings, a perfect expression of the ideal city which later fascinated Renaissance urban planners (Lugli, 2006:7). From remote times, a circuit of temples and sacred groves formed a ring some five Roman miles from the city center, understood as either the Capitol, or the Golden Milestone located on the western extremity of the Roman forum. The radius was equivalent to ten kilometers in modern reckoning, and defined the limits of the Roman agro, so important for food provision in the republican city. The urban area proper extended some three kilometers from the central milestone, constituting a municipal tax zone for city residents (Lugli, 2006:11). From the classical antique period this three-kilometer radius would have defined the city proper, including the suburbs. Beyond came the Roman agro, encircled in turn by the Roman campagna.[1] Piero Maria was the son of a noted archaeologist and topographer, and attempted to incorporate these ideas in the 1962 Rome master plan which he helped shape. He was one of the five experts called upon to draft the plan, the first to include all of Rome's vast territory. The plan, however, was largely ignored in the process of growth it was intended to order.

The dominant feature of the ancient Roman city is the orthogonal plan. It is thus a surprise to discover in Rome the importance of the ring. The Five Mile circuit has had persisting relevance well into contemporary times. For most of its history after the fall of the empire, the Roman periphery was abandoned swampland, excluded from any remotely urban condition. It was then reclaimed, acquiring agricultural importance. In rapid historical succession it evolved into what we think of today as the urban periphery. Reclamation was a prerequisite for urban growth, and a variety of initiatives were adopted in different historical periods to turn swampland into arable fields. One of the most important of these was the Petrocchi Law of 1878, which established special provisions for the reclamation of land falling more or less within the Five Mile circuit (Valenti, 1984:212). The same circuit defined the limits of regulated urban growth in the 1931 master plan, which were preserved in the 1942 variation on the plan. Today the circuit is an important urban marker, corresponding roughly to the great ring road (GRA or Grande Raccordo Anulare) built originally in the 1940s under the direction of the engineer Carlo Gra. Today, with the city's sprawling growth, residence inside the GRA is a sign of being part of the consolidated city.

Background to the modern periphery

The contrast between the urban and the rural is another 'great dichotomy' which shapes urban life in Italy even today (Bobbio, 2006). Typically, this is a contrast between inclusion and exclusion, although from ancient times country life has been celebrated in a bucolic vision exalted by the city's civilizing action (Sereni, 1997:40). The prosperity of all Italian cities – most notably Siena – was historically linked to that of the countryside. The linkage was less evident in Rome, owing to its prominence as home to the Catholic Church and its vast size. Rome, with a territory of 1,287 square kilometers, is the most extensive city in Europe.

From ancient times up to the present day the relationship between the center of Rome and what can be considered a changing periphery has been eminently unstable. The periphery, understood as an assemblage, has always blended together spatial relations, power and narrative. As we have noted, through changing configurations this assemblage is most easily understood as exhibiting a concentric ring form. In the celebrated apex of the classical city's history – the times of Augustus – the center was the area around the forum, home to boundless privilege. Located just outside this ring, especially to the east of the city, was what then would have been the underclass district. This was the suburra, whose outer limit was contained by city walls dating to the early centuries of Rome's history. Outside these walls was a belt of suburban villas, called horti, belonging to Rome's elite. The great patron of the arts, Maecenas, owned a villa just outside the walls on the eastern fringe of the city. This was the landscape of poetry, the bucolic vision expressed by Virgil in his *Georgics*; a vision with lasting impact on later poetry and painting. The swath of luxurious suburban villas extended for some three kilometers, and even beyond, especially to the east. Remoter villas were proper farms combining sophisticated owners' quarters with the functional units of farm workers (Lugli, 2006).

This bucolic life was destined to change rapidly with invasions dating to the 5th century AD. After these events decline was the trend, articulated through countless vicissitudes persisting over the duration of a complex history, including brief revival in the early middle ages (Carocci and Vendittelli, 2004). It was in these circumstances that the greater part of both the Roman agro and campagna was abandoned. Yet the ring configuration persisted with new topographical meaning. Even parts of the historic city center were abandoned. The eastern high ground of the city proper – as defined by the 18-kilometer circuit of the Aurelian walls – was entirely abandoned after

devastating wars and invasions. By the late Renaissance, this area became
known as the 'uninhabited' (Krautheimer, 1980:68). Just as it was becoming known for its abandon, so this part of the city attracted the interest of
the rich, soon being repopulated with sumptuous suburban villas. The ring
outside the Aurelian walls was instead comprised of vineyards and farmland used to produce food for the city. Beyond, from the high middle ages
to the time Rome became the nation's capital in 1871, little could be found
other than isolated rural settlements. Encircling these fringes of the Roman
campagna was a series of consolidated hill towns, including the Castelli
Romani to the south, Tivoli to the east, and the towns approaching Viterbo
to the north. Between the circle of these hill towns and the ring around the
Aurelian walls abandon prevailed, interrupted by scattered pastoral pursuits and limited agricultural activities (Lugli, 2006). The campagna was
divided into vast estates belonging to a handful of Roman nobility, land
so wild that it hardly warranted cultivation. Wheat was produced on some
estates, but stable agriculture was hindered by labor scarcity. Agricultural
workers came down on a seasonal basis from the surrounding hills, even
from as far away as Abruzzo. Their temporary summer dwellings consisted of precarious shacks organized in what could hardly be considered
settlements (Martirano and Medici, 1988). The grazing of sheep and large
livestock like cattle and buffalo was prevalent, and the Italian cowboy of
the campagna was celebrated in painting as late as the early 20th century
(Valenti, 1984). Even today sheep grazing persists in parts of the periphery,
and was prominent in images of peripheral Rome well into the 1960s, as in
scenes recorded in Pasolini's film *La ricotta*.

From the late Renaissance onward, the Roman campagna resonated
with semantic dissonance. On the one hand it was celebrated as preserving
a vision of lost ancient times. On the other, these vast tracts of unproductive land constituted a public health concern. Balanced between romantic
memory and the quest for progress, mild reform efforts were adopted by the
papal authorities. The most significant was devised by Pope Alexander VII,
who mandated the first modern land register of the Roman campagna, the
Alexandrian cadaster of 1660 (Passigli, 2012). The register was intended to
raise taxes on rural estates to pay for vital infrastructure, like roads and
drainage systems. It also required that landowners produce food for the
city. The illustrated maps that comprise the land register show the circuit
of the Aurelian walls surrounded by countryside, with pathways, orchards
and vineyards; now almost entirely engulfed by urban growth. The efforts
of Pope Alexander, like those of his ecclesiastic followers, had little effect,
and by the early 1800s the Roman periphery was still as desolate as before.
Writers of the day blamed the failure to develop on landowners who constituted an antimodern elite wishing to preserve the wild and romantic character of their estates (Valenti, 1984).

In the 17th century this wild abandon became a template for European
landscape painting. The image of the campagna gained European currency

with Claude Lorrain, a French painter who first worked in Rome as a pastry chef and went on to become Europe's greatest landscape painter (Wine, 1994:115). His method was based on the minute documentation of the natural heritage of the campagna, providing elements that were combined into a highly romanticized vision of a pristine past surviving into a contemporary period. In Claude's understanding, the campagna formed a ring around the city, starting with parts of Monte Mario which were no farther than two kilometers from the Aurelian walls. The campagna became 'a perfect image of the state of innocence,' completely undomesticated and for that reason pure (Wine, 1994:29). His patrons were conservative estate owners who would hang his paintings on walls overlooking their property, so they could use his romantic standard as an aesthetic measure of the natural landscape that was visible to them through the windows in the walls where the paintings were hung.

Even the abandoned cityscape inspired a romantic vision, which writers such as James Hawthorn in 1860 described as worthy of being painted precisely because it represented a 'picturesque' image harking to an archaic past (Strappa, 1989:35). It was thanks to the British influence that Italians came to use the term 'pittoresco' with no slight irony, as an image of beauty justifying Italy's backward condition (Gisotti, 2008). Yet other grand tourists were appalled by the bleak conditions of the Roman hinterland. Charles de Brosses referred to the Roman campagna as 'intensely sad and hideous' after his tour of 1739 (Beaven, 2014:79). He was of course familiar with Claude's idealized vision, and for this reason all the more vehement in his judgment. Gilbert Burnet, travelling in 1680, noted the contrast between actual and potential conditions of the land, stating that notwithstanding its great agricultural potential, the campagna was one of the most 'dispeopled' areas of Italy (Beaven, 2014:81). Protestant travelers were delighted to use these bleak conditions as an opportunity to express negative judgment against the Catholic Church. Another writer of 1740 spoke of 'a horrid thing called the malaria,' long considered a national calamity (Beaven, 2014:85). Indeed, it was precisely this scourge that spurred prodigious reclamation efforts after the unification of Italy and the termination of papal rule. An American traveler wrote in 1810 that he had never known the opposite of human progress until he visited the areas around Rome (Felisini, 2009:299). In 1881 there were only some 12,700 people living in the Roman campagna, housed largely in seasonal shelters made of corn stalks, straw and mud (Celli, 1984:237). This was the price exacted by the romantic vision.

Only gradually was progress imposed. However picturesque, the wild swamplands were a health hazard and an impediment to development. Land reclamation was the solution to this millennial scourge, and could increase property value even 50-fold. But the financial cost of reclamation was exorbitant, often exceeding the benefits accrued through property improvement (Cavallo, 2011). The Baccarini law of 1882 spurred transformation, defining reclamation as an activity of 'fundamental public utility.'

This law mandated the creation of consortia of landowners to oversee the drainage works, called Consorzi di Bonifica (Cavallo, 2011:33). At least one such company is still present on the eastern periphery of Rome, with farm buildings now publicly owned yet still known by the name of the historic reclamation company, ALBA (Anonima Laziale Bonifica Agraria).

The Baccarini law had effect all over Italy, but Rome was a special case, and in 1910 a law was passed with specific pertinence to the Roman agro.[2] The 1910 law mandated the creation of rural communities outside a radius of 5 kilometers from the city center. The mandated settlements would repopulate the landscape. These were to be stable communities, not the precarious dwellings of seasonal workers. Each community so defined was to have a population of no fewer than 25 families living in houses with adequate sanitary standards, surrounded by agricultural property on a plot of at least half a hectare. From an etymological standpoint it is vital to note that these communities were termed borgate, the same term used to designate the fascists satellite communities sanctioned in 1931. Special funding through long-term mortgages was available to finance these initiatives, and as an added incentive the borgate thus created were exempt from taxation for 20 years. If the traditional antimodern elite landowners were unwilling to take the initiative, the property could be expropriated by entrepreneurs who could then construct the new settlements. A national commission (Commissione di Vigilanza per l'Agro Romano) was instituted to administer requests for expropriation, and a municipal department in Rome was created to oversee the founding of the new communities. This was the Ufficio Agro Romano responsible for halting illegal building, and overseeing urbanization works which complemented the reclamation.[3] Through a mixture of encouragement and coercion the antimodern elite were at last forced to relinquish their deleterious privilege (Cavallo, 2011:116).

This sequence of events established the conditions for urban growth. Drainage was fundamental, but so was the leveling of a terrain comprised of abrupt hills and steep inclines. In its wild state the Roman campagna was far too irregular for agricultural work. Leveling was achieved with heavy equipment funded up to the mid-1930s by state-secured mortgages (Martirano and Medici, 1988:290–298). The most impressive of these interventions led to the creation of a sizeable agricultural estate called Maccarese. Private entrepreneurs, starting their activities in 1925, created this farm of some 3,200 hectares on what had previously been unproductive land. It is today publicly owned.

The urban periphery thus did not grow out from the city to occupy land that had no previous function, a tabula rasa or semantic void (Turri, 2014). Instead, growth involved a complex interplay of multiple actors in which rural changes played a critical role, leaving traces which are still evident today. The epic permutation of the periphery paralleled massive transformations in the urban environment, and as always change engendered controversial interpretations. In the 1890s the great Italian poet, Gabriele

D'Annunzio, described Rome's perimetral expansion as an immense gray tumor which grew from the flanks of the ancient city, sapping its life in the process (Strappa, 1989:35).

United Italy

Rome's master plan of 1883 was limited to the area located within the Aurelian walls. Regulation of the outside perimeter had to wait until 1909. In the intervening period, the urban fringe became a site for informal growth. The terminology of popular parlance used to classify these spontaneous settlements borrowed from events far removed from Rome. The 1880s witnessed Italy's first effort to establish a colony in what was then Abyssinia, events which captured the headlines of the Italian press. The peripheral communities established in those years around the Aurelian walls thus came to be known as Abyssinian villages, or as more positively connoted borghetti, which we can translate as hamlets (Sanfilippo, 1993:331). They were comprised of shacks built with informal materials, home in all of Rome to some 10,000 residents (Martinelli, 1986:20). The settlement at Porta Metronia, just outside the walls on Via Appia, was comprised of 32 shacks each containing 9–20 families. Most of them had basic water and toilets, beaten earth roads and shared laundry facilities. The Porta Metronia shacks were torn down in 1938, and the families transferred to new fascist borgate. Another famous Abyssinian village was outside of Porta Cavalleggeri, on the western fringe of the city in what the 1660 land map represents as vineyards. The spontaneous village was demolished in the 1909 plan, and replaced by solid midrise buildings which are still present today (Sanfilippo, 1993). The original inhabitants of the Abyssinian village established a new community about five hundred meters to the west, along what was once Via Aurelia. This is Borghetto di Monte del Gallo, today an attractive low-density residential district comprised mostly of modest midrise buildings. A scattering of Abyssinian shacks which survived the 1909 relocation were torn down in anticipation of the 1960 Olympics, and the survivors were transferred to public housing at Casetta Mattei on the far western fringe of the city. Abyssinian villages were part of a new periphery generated by massive urban growth. The entire city was transformed in this upheaval, including areas inside the historic walls.

In following the city's growth, we must note that the center of Rome is divided into local administrative districts known as rioni. There were 14 rioni in 1586. Esquilino was added in 1874, and another seven in 1921 (Zeppegno, 1996). The total number today is 22. Martinelli (1964) furnishes a careful demographic study of these changes, showing a decisive contrast between the central rioni, located along the Tiber, and the peripheral ones, mostly newer residential areas in contact with the Aurelian walls. Between 1870 and 1910 the peripheral rioni – with their broad avenues and impressive coordinated façades – acquired population, while the historic central districts lost residents. The driver of change was the poor quality of inner-city housing, measured in terms of the frequent absence of running water, toilets and bathrooms (Martinelli, 1964:141). Beyond the peripheral rioni were the new quartieri, forming a ring outside the Aurelian walls. While the new buildings in the quartieri were intended mostly for government workers, there were also pockets of poverty. Some 10% of homes in the prestigious Ludovisi quarter in 1951 were without running water. In the same census year almost half of the homes in the rioni lacked an internal bathroom (Martinelli, 1964:141). Thus, in Pasolini's day the central areas of Rome were characterized by substantial underprivilege, surrounded by two rings of privilege, in turn encircled by the vast area of the Roman campagna then undergoing a process of radical transformation.

Agnew documents the massive alteration of the historical center following 1871 (Agnew, 1995:32–39). On the one hand was an extensive program to create clearances for a new street system, and new public spaces. On the other, vast areas within the Aurelian walls were earmarked for the building of new residential districts. All of this occurred with the weakest regulation, notwithstanding the 1883 masterplan. Expansion outside the walls was only regulated after 1909, two years before the 1911 World's Fair that celebrated the anniversary of national unification. It was in these circumstances that the Abyssinian villages were demolished. The progressive mayor of the time, Ernesto Nathan, showcased the nation's capital, cleaning up the eyesores which surrounded the walls. Nathan also municipalized the city's public transportation system, and founded the still functioning public company called ATAC. The most characteristic feature of the transport system in those days was the circular tram line that ran in both directions just outside the Aurelian walls. The construction of this line required that the perimetral areas be freed of informal accretions. Trams transited around the wall circuit, connecting the northern extremity of the Aurelian walls to the southern one using tracks that ran down the left bank of the Tiber embankment road. These tracks were torn out in the late 1950s (Zannoni, 2010). Nathan was also the first mayor to dispose of a national fund for public housing, mandated in the Luttazzi law of 1903. This led to the creation of a public housing authority, called ICP (Istituto Case Popolari), replaced in the 1990s by a regional authority called ATER. The first public project funded under this law was at San Saba, a wonderfully designed community just outside the city walls not far from Via Ostiense.

Mixed urban growth leading up to the fascist period

Rome's first public housing dates to 1883, built in the area of Testaccio using municipal funds. Historically Testaccio was an open field known as the Prati del Popolo, a gathering place for celebrations organized among the popular classes (Mura, 2014). This was also the intended location of an early industrial district, and the site of the municipal slaughterhouse. After

1909 this western corridor expanded into the industrial district built along Via Ostiense, adjacent to which the garden city of Garbatella was founded in 1920.

Discussion concerning public housing started almost as soon as Rome became the nation's capital. The municipal commission charged with exploring public housing policy issued in 1871 a recommendation against creating homogenous working-class neighborhoods (Accasto et al., 1971:155). This recommendation was drafted two months after the Paris commune uprising, suggesting that it was dictated by reasons of political expediency rather than humanitarian concern. The mandate to create mixed income areas was later adopted as a cornerstone in Italian urban planning. It is not uncharacteristic that the recommendation was ignored in the short run, and Testaccio was built as an exclusively working-class district. In 1888 two other working-class neighborhoods on a smaller scale were created, also with local funds, one at San Lorenzo and the other at Santa Croce. Both were located in proximity to Rome's major train station, in what were identified as axes of industrial growth (Insolera, 1976:74–75). More substantial public housing was constructed after the 1903 legislation. In addition to San Saba, and the garden cities of Garbatella and Aniene, a concentration of early public housing was built in the area north of the Vatican. This is Prati della Vittoria, a well-connected central area where land had been expropriated by the city for the 1911 World's Fair. Afterward it became available for other public functions, in this case public housing. Needless to say, these units today are extremely attractive because of their central location and high-quality architecture. This was the beginning of Rome's formal peripheral growth, which was later consolidated with the 12 fascist borgate mandated in 1931. All 12 of these were built at a significant remove from the historic city. The fascist government continued to gut out parts of the historic center, creating ever new need for housing which was magnified by the continual flow of new residents from rural areas. The fascist planners also projected growth along a western corridor that ran from Ostiense, out through the new district of EUR, and on to the port town of Ostia.

Today fascist urban planning is remembered especially for its borgate. The efforts taken in this period to disseminate middle-class housing throughout the periphery are instead overlooked. Plans to create a mixed income periphery start with the role played by villini in 1909. One of the most desirable residential typologies in Rome is the small one- or two-family home locally termed a villino. These are heirs to the ancient villa with its long and noble tradition. The master plan of 1909 made provisions for entire neighborhoods of villini, such as Monteverde, just outside the Aurelian walls on the southwestern side of the city. Villini were intended to house middle-class families, but at times were used as worker housing, such as found with railroad-worker housing built in different parts of the city.

Another type of building connoting status is the palazzo, with its palatial etymology. The term is used interchangeably with palazzina. This represents an ideal type in Rome and many Mediterranean areas (Agnew, 1995:138). These buildings are typically developer built, but one also finds self-built homes adopting this typology, an expression of what is sometimes called imposed illegality ('abusivismo per necessità'). If one wanders through the Roman periphery one will find even four-story eight-unit palazzine which have the same surname on the intercom for every flat. This may be the case of a migrant who in the 1950s built a palazzo large enough to have one unit for each of a numerous offspring.

Fascist efforts to distribute middle-income families throughout the city are especially associated with the palazzina. In the 1920s this typology took the place of the villino, so prominent in the 1909 plan, in a policy measure designed to densify areas chiefly intended for middle-income housing. Rome's population continued to grow up until the 1981 census year, and there has always been heavy demand for housing. Housing pressure was especially strong in the 1920s, and limited resources available to satisfy the need. It was thus that the government passed a law in 1920 providing a legal definition of palazzine, and allowing villini to be transformed into this new building type (Muntoni, 2006). The law defined a palazzina as having a maximum of five stories, including optional ground floor commercial space. It also specified generous site requirements, so that palazzine would be surrounded by what was effectively a condominium garden. The number of units per palazzina was limited to no more than 12. They were of higher density than the villino, yet still expressed an intimate family environment in a densification process funded by private developers.

Alessandra Muntoni (2006) devotes an unusual study to this building type, showing its surprising pertinence in the modern evolution of the city. Today the palazzina is a byword for middle-class respectability. Muntoni notes that today's progressive press is hostile to this building type because it suggests distance from a broader shared community, a retreat into the privacy of the family. Being the signature fascist building also plays a role. Muntoni reminds us that the builder of a palazzina is known today by the term palazzinaro, a word used to deprecate real estate developers. A palazzinaro is thought of as a 'speculator' who implicitly pursues private advantage over public welfare. Yet hostility notwithstanding, the palazzina is the typical building of this city. Insolera describes the importance of the palazzina, noting that it is the housing type privileged by Rome's bourgeoisie. 'Given that Rome has always been essentially a middle-class city, we can state that Rome is to a great extent a city of palazzine' (Insolera, 1976:96). Muntoni (2006:144) reproduces a map showing the extensive areas devoted to palazzine in the 1931 fascist plan, spread throughout the city's urban fabric. Far from being a marginal element, the middle class has deep and lasting roots in the Roman periphery.

Monteverde

Monteverde provides a convenient example of Rome's organization. It was home to Pasolini between 1953 and 1964, first in via Fonteiana 86 and later in Via Carini 45.[4] The latter location is a more central part of Monteverde.

At both locations Pasolini lived in respectable palazzine. Monteverde was planned as a residential district in 1909, designated mostly for villini which were built by cooperatives representing state employees (Dal Mas, 2006:190). While one still finds villini today, many were replaced by the palazzine allowed by the 1920 law. In the 1909 plan, the villini were located mostly along the eastern part of Monteverde, approaching the Aurelian Walls. Off to the west, the 1931 plan permitted the building of two intensive residential blocks. They occupy a prominent position on the crest of the hill which is Monteverde, visible throughout the city's southwestern quadrant. Close to these intensive buildings is another dense housing project, but of a smaller scale, constructed under a plan instituted by Mussolini in 1924 to supply housing to government workers, called INCIS housing. We thus find in Monteverde by the 1930s a mixture of housing types, some public housing for middle-income government employees, some intensive middle-income housing built by developers, and villini from 1909 (Dal Mas, 2006).

The villini are the core of what is today called Monteverde Vecchio to distinguish it from Monteverde Nuovo extending out to the west. The new part of Monteverde was mostly farmland in the 1930s, speculatively developed with palazzine in the 1950s and 1960s. As is often the case, planning was weak and country roads were simply broadened to create a street network which barely serves the needs of the modern city. The 1909 planning map shows farmland in this area, with a discernable S-shaped country road corresponding to today's Via Fonteiana. This was straightened out to form an inverted V-shape, today surrounded by intense urban growth. Just above this V we find Pasolini's palazzina of 1953. A little further down, Via Fonteiana intersects Via Donna Olimpia, at a point which was open countryside in 1909. In the early part of the 20th century this depression was called the Reed Valley, owing to the wild reeds that grew there, as they do in other parts of Rome's abandon countryside (Dal Mas, 2006). The area's lack of utility probably related to flooding caused by rain runoff channeled down the path of Via Fonteiana from the northeast, and Via Donna Olimpia from the northwest. As low-cost land, it was the ideal site for a public housing project that was established there in 1931. Although unrecorded in any published document, the site would have required drainage works before the project could be built.

The isolated and chiefly rural area around Via Donna Olimpia acquired importance after the building of a bridge in 1929 connecting Monteverde Nuovo to a boulevard, envisioned in the 1909 plan, which led to Viale Trastevere in one direction and San Camillo Hospital in the other. From San Camillo this road eventually connects to the path of Via Portuense, leading all the way out to the coast. In the fascist period this western corridor acquired new urban importance with the planning of the city's growth out in this direction (Annappo, 2003). It was in these circumstances that the public housing project was conceived. It was originally known as Pamphili I, taking its name from the papal family that had once owned property in the area. The architect was Innocenzo Sabbatini, a respected professional with experience in designing Rome's garden cities in the early 1920s (Villani, 2012:92). The Pamphili project consisted of three nine-story buildings, so designed as to render in plan view the three letters that comprise the name by which Mussolini was known to Italians, DUX. Among their many desirable features, these buildings were characterized as being well ventilated, in contrast to the windowless shacks that housed the urban poor (Villani, 2012). Typically, the flats were comprised of a single living room, a bedroom and basic toilet facilities. Rent was 50 lire for a one- or two-room flat, 135 lire for a four-bedroom flat. 540 flats were built, and 300 were assigned by lots to aspiring tenants on a long waiting list. Some 50 homes were given to families evicted from the historic center. A maximum income limitation qualified families for assignment. Yet families were also required to document enough income to ensure payment of the rent (Villani, 2012:94). The families were poor but not destitute.

Innocenzo Sabbatini was an advocate of the Roman school of architecture, a school which wished to restore the sense of enclosure they believed had been lost in the city's 19th century expansion (Regni and Sennato, 1982). In the effort to create continuity with Rome's baroque and medieval areas, the school documented systematically the historic city center, parts of which were being destroyed by fascist urban planning (Stabile, 2012). Today, the Pamphili project preserves its original character with a high architectural standard. It exhibits subtle decorative patterns, and an intriguing plan consisting of a well-conceived range of geometric forms, with courtyards of varying dimensions and under-passages that reveal changing perspectives on an urban microcosm echoing the inspiration of the historic center.[5] The figure ground inspiration is patent.

Another public housing project featured in Pasolini's novels, and possessing similar celebrated merits, is the INA project called Tiburtino IV. It is located in the eastern part of the city. The INA plan was mandated in 1949, and represents a moment of excellence in Italian public housing (Guccione et al., 2002). The signature INA project is precisely Tiburtino IV, designed by Ludovico Quaroni and Mario Ridolfi. Both architects were tied to the PCI, and their project won broad critical acclaim. Like Pamphili, Tiburtina IV has a wonderful sense of enclosure, with a central piazza reminiscent of medieval towns like Siena or Lucca. Though completed as part of a single plan in a two-year construction period, it was designed in such a way as to convey the impression of successive architectural interventions. In Colin Rowe's terms, it wished to imitate a collage rather than a utopian design, and was successful in the effort. This project is located not far from Pasolini's home in Rebibbia, and features prominently in *A violent life*.

In both of his novels Pasolini singles out expressions of architectural excellence as critical narrative settings. As we have noted, the Pamphili project is no more than a five-minute walk from Pasolini's 1953 home. In Friuli Pasolini boasted that within a ten-minute bicycle ride he could pass from one linguistic area to another (quoted in Rhodes, 2007:17). In Rome only 250 meters from the comforts of his bourgeois home he could find an expression of the desperate periphery which provided the setting for his novels and films.

Two 1931 borgate: Quarticciolo and Trullo

The borgata that received most of Ferrarotti's attention in the 1970 volume was Quarticciolo[6] on the eastern fringes of the city. This borgata was designed by Roberto Nicolini, who by historical accident was the father of one of the symbols of the 1976 PCI administration, Renato Nicolini. Renato was a champion of the Roman periphery, and invented a series of summer activities (L'Estate Romana) aiming to add value to Rome's public areas, both in the center and the periphery (Nicolini, 2011). Ferrarotti paints Quarticciolo in harsh terms, owing not only to its social conditions – which were indeed difficult – but also to its physical setting. He claims that Quarticciolo represents an image of desolation: 'the uniformity of types, the dull surfaces and dusty courtyards define a squalid, disconsolate environment' (Ferrarotti, 1979:80). In the same pages he describes the absence of any form of community, with strong individualism and universal diffidence. We are presented with a dismal physical setting, as well as an atomistic social environment of the type described by E.C. Banfield (1958) in his negatively biased description of the Italian south.

It is easy to refute Ferrarotti's characterization of Quarticciolo's physical environment. Roberto Nicolini was a distinguished architect, the child of a noted Sicilian sculptor and father in turn to Renato Nicolini, an architect, professor and noted left-wing politician. Renato appreciated the high architectural standard of Quarticciolo, and wrote a contribution to a catalogue which stressed its merits (Nicolini, 2010). Renato's essay faces the challenge of distancing the author from his father's politics, while expressing appreciation for the merits of the architect, and personal affection for his father. Not all architects working under fascism were fascists, but it is clear from Renato's treatment that his father had sympathy for the regime. While rejecting his father's political views, Renato stresses the success of Quarticciolo in terms of its design. Far from being dull and uniform, it successfully blends together classical orthogonal urban design with the celebrated medieval tradition characterized by a human scale, a balanced sense of enclosure and the clear definition of monumental areas intended to celebrate a civic spirit (Fig. 2.1). Quarticciolo's long central spine was designed in the rationalist tradition, with lateral buildings defining a series of open squares. Dominating the complex is a town bell tower – campanile – of medieval inspiration. The well-defined community space centers on a service area once containing the daily produce market. As Renato notes, the overall impression is almost metaphysical, a rustic version of a rationalist city.

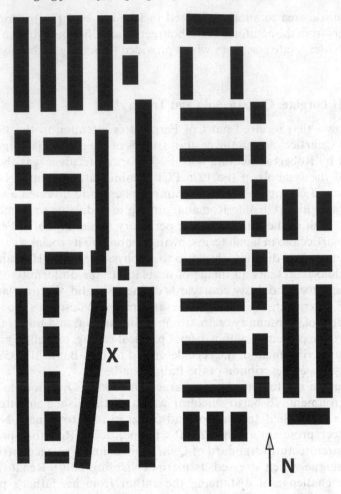

Figure 2.1 Figure ground representation of Quarticciolo. The letter X shows the position of the theater visible in Fig. 2.2. This diagram reveals a strong sense of order and enclosure (© Gregory Smith.)

The inner squares were intended to house family gardens, and indeed today are maintained informally by the residents of the housing complex. Tenants today voice the familiar complaint about the poor maintenance provided by the public housing authority (ATER).[7] But at the same time, one can discern the care residents expend in maintaining their shared spaces. 2007 was a turning point for the neighborhood, when a library and 150-seat theater were inaugurated taking the place of the original market (Fig. 2.2). The library is part of an extensive network managed by the City of Rome, frequented especially by university students who spend much of their time studying at this facility. It is a lively community hub equipped with a coffee

Figure 2.2 Quarticciolo library and theater. (Photograph taken in August 2020 © Gregory Smith.)

shop and restaurant. The theater is likewise highly successful, and since 2013 has been part of the City of Rome's theater network promoting cultural activities all over the city, including the urban periphery.[8]

Pasolini would have been perfectly familiar with this project, owing to his explorations of the periphery, and the fact of having been hired to play a supporting role in the feature film *Il Gobbo* (*The Hunchback*) (Lizzani, 1960). He appears in the role of a pimp, with his voiced dubbed over to give an authentic Roman flare. In addition to being one of Rome's most noted borgate, Quarticciolo is still famous for the bandit turned resistance fighter known as the Hunchback. Today if one asks local residents about the Hunchback, they describe an ambivalent figure, a socially disturbing criminal, but also a resistance fighter who supported the community. The theme of the 1960 film resonates with the community's own self-image, torn between solidarity and alienation. Even a conversation at the local coffeeshop will show that many citizens believe a community has always existed at Quarticciolo, united by the daily struggle of attempting to achieve better conditions of life. Pride in place is also reflected in the library's collection of volumes on the history of Quarticciolo.

Roberto Nicolini was a prolific architect, and a major force behind another noted borgata called Trullo. This was commissioned in 1939 to accommodate Italians who were returning from abroad in anticipation of disruption caused by the Second World War (Venditti, 1989). The project also housed citizens displaced from the city center by fascist demolitions. Trullo is located along a peripheral road called Via Affogalasino ('Drowned Donkey'), a name which chronicles the area's historic hazardous conditions. The name was adopted about 1500 to designate a drainage channel later replaced by the modern road (Venditti, 1989:37). The current Via Trullo takes its name from a toponym appearing in maps in the early 1800s (Venditti, 1989:22). The long-standing stigma as abandoned swampland carried directly over to the housing project, and in Rome the word Trullo persists as a byword for urban marginality. The area was first reclaimed by an industrial group in the early 1900s to build a weapons factory. Shanties sprang up in proximity of the industrial site, then cleared away when the borgata was built. Trullo was also designed by Roberto Nicolini with a similar blend of orthogonal and medieval styles, and a grand central square dominated by an impressive church. It also has a theater, which today enjoys considerable success, with schoolchildren bused in from all over the Roman periphery for matinee performances. In recent years the entire neighborhood has been involved in an extraordinary revival led by the activities of street poets and street artists.

Trullo is one of two historic borgate located on the right bank of the Tiber River. The other is Primavalle, also exhibiting a rationalist style. Most historic borgate are instead located along the main consular roads on the eastern side of the city: Val Melaina, Tufello, San Basilio, Pietralata, Tiburtino III, Prenestino, Quarticciolo, Gordiani and Tor Marancia. This count leaves out Acilia, the most remote from the center, at about 15 kilometers from the Golden Milestone. Acilia sprang into being owing to the opening in 1928 of a motor-way connecting the center of Rome to the port of Ostia, following the path of the ancient Via Ostiense.

Via Ostiense follows the course of the Tiber along its left bank. The main consular road on the right bank is Via Portuense, taking its name from Rome's port. Together they form a corridor leading out to the Tiber estuary. The roads enclose a low-lying floodplain which late in its history was given over to agriculture. Via Ostiense starts at Porta San Paolo, just outside of Testaccio, and is home to the industrial district we have described and close to the garden city of Garbatella. The area along Via Portuense is more complex, both in topographical and urbanistic terms. As it heads out from the city center, Via Portuense traverses the southern extremity of a catchment area starting on the western side of the Janiculum hill, in proximity of Monteverde, and enclosed by reliefs off to the west. Rain runoff collects north of Primavalle, and carries down to the Tiber at a point known as Magliana.[9]

Via Portuense starts at Porta Portese on the Aurelian Walls, and runs in a fairly straight line out to the port. An important offshoot of this road is Via della Magliana, which hugs the northern course of the Tiber before joining up again with Via Portuense. At a point not far from Porta Portese, Via della Magliana follows a straight westerly path while the Tiber bends to the southeast. The enclosure thus formed between Via della Magliana and the Tiber River houses Rome's highest density neighborhood. This is Magliana, giving its name to Rome's famous criminal organization called the Banda della Magliana, celebrated in novels and films (Renga, 2017).

The area traversed by Via Portuense has much to say about Rome's complex growth. Slightly beyond the Pamphili project we find an intensively developed area dating to the 1950s, where palazzine are the chief building type. As we head south down Via dei Colli Portuensi, we encounter a neighborhood comprised chiefly of high-income gated communities. Beyond this is Via del Trullo, with its housing project on the floodplain and an abundance of informal housing on the hills to the west. Two kilometers later we reach Corviale, Rome's most famous public housing project. It dominates an area containing a core of informally built homes along Via Casetta Mattei. Via Casetta Mattei also has public housing projects dating to the 1960s where, among others, some residents of the Porta Cavalleggeri Abyssinian village were relocated.

Corviale is a building one kilometer long, a linear construction done in the best modernist style. It is part of a PEEP, a complex territorial mechanism which guided urban growth all over Italy, created by Law 167 of 1962. PEEPs take their name from the acronym for the plans which governed urban growth in the 1960s and 1970s, Piani di Edilizia Economica e Popolare. Rome has 54 such plans, and one is precisely Corviale.[10] Most PEEPs contain a core of public housing, surrounded by cooperative housing intended for middle-income families, and areas left free for higher income development. The kilometer-long building was designed by a distinguished Roman architect, and member of the panel of experts that drew up the city's 1962 plan, Mario Fiorentino. The building, whose construction started in 1972, is immensely famous all over Rome where it is referred to as the big snake ('serpentone'). As a modernist construction it rejects any idea of a psychological relationship between the city dweller and its monumental mass. In environmental terms the building is supremely efficient, housing 1,200 families on a compact site. The building has received abundant media attention as a symbol of the modern periphery. Right-wing politicians have called for its demolition, not least because Fiorentino was aligned with the PCI (Casalini, 2017). Yet progressive forces advocate its preservation, supported by citizens, rap groups and street artists. The progressive forces have prevailed, and the big snake is currently undergoing major regeneration, enhancing the original design and integrating it better into its natural setting.[11]

A crucial symbolic year for architecture and urban planning was precisely 1972, the year Pruit Igoe was demolished. This event in some ways marks the end of modernist utopian aspirations (Jencks, 1977). The demolition also embodies a noted fallacy in planning logic which blames architecture for urban dysfunctions which have social and economic causes (Freidrichs, 2011). It is certain that the destruction of Corviale would not solve the problem of marginality. It is equally certain that this option would entail massive cost and present formidable environmental challenges. Though from the outside Corviale can be seen as an example of brutalist architecture, many citizens living in the project are impressed to be part of an extraordinary modernist facility. They ask for better maintenance, better public transportation, more social services (Martini and Parasacchi, 2004). This is the course being pursued today.

Urban informality

The periphery is far from homogenous, containing a maze of forces assembled over such a long and complex historical process as to be difficult to disentangle. A pertinent example of the transformation of the romantic campagna into farmland and then into a peripheral urban area is Bravetta Pisana. This is on the western side of the city, part of the catchment basin described above. Bravetta and Pisana are the names of two historic country roads that enclose an elevation today designated by local residents using a combination of the two names. Maps from the 1950s reference the hilltop community of this variegated neighborhood as Borgata dei Villini.[12] The central core of the community was almost certainly developed thanks to public support inspired by the 1882 Baccarini Law. Indeed, a massive publication collecting all the state-secured mortgages granted to aid land reclamation in Rome reports a series of such mortgages issued to entrepreneurs in Bravetta starting in 1906 (Eramo, 2008:175). Contemporary residents remember this as having been an agricultural district up until the 1950s, producing vegetable products sold in Rome's extensive neighborhood market network. The gardens then gave way to speculative urban growth.

What is intriguing about Bravetta Pisana is the distinction between the northern and the southern parts of the neighborhood.[13] The northern part seems to have grown thanks to the provisions of reclamation laws wishing to create stable rural communities in the Roman periphery. As we know, each building was surrounded by a generous plot allowing for the presence of a perimeter garden. In a later period, the generous proportions of the original lots made it possible for homeowners to redevelop their property by building palazzine. This redevelopment of existing sites appears to have occurred more than once, with many restructuring efforts dating to the 1980s. In this northern part of the neighborhood today one finds a sharp contrast between modern palazzine of a high standard built in the last 30 years, and palazzine of an older date which are in a significant state of disrepair. Local residents support the view that the maintenance of some buildings is neglected because the owners hope they can speculatively redevelop their sites, constructing larger units with greater market value.

The built part of southern Bravetta Pisana dates instead to a later period, the 1950s. Here land was parceled out informally with few zoning provisions, as is so common in Rome's periphery. The roads are narrow, with small and erratically-shaped lots. There is no overall logic to the road structure, and the main street of the neighborhood is a dead end. Many of the side streets are dead ends as well. Residents cope with this poor structure by using informal pathways that connect the different parts of the neighborhood. The homes were mostly informally built. The setback from the streets is entirely irregular, dictated by the whim of the individual builder. Yet it is paradoxical that these informal units in the southern part of the neighborhood are uniformly well maintained. All the homes in the southern area appear to be inhabited by the original tenants or their descendants, mostly rural migrants who arrived in great waves in the 1950s. These sites are not being speculatively redeveloped apparently because of the limitation of small plot size. The larger plots in the north date to the area's history as an agricultural borgata. The southern lots are instead a product of the later informal parceling out of land with few or no formal guidelines. This contrast shows the continuing importance of historical trajectories in the city's growth.

There is increasing awareness of the importance of informal practices in urban planning, although mostly focused on developing countries (Roy, 2005). Recent attention has been addressed to this issue in Italy, showing that these practices yield mostly negative outcomes (Berruti, 2019). Roy notes that the distinction is often not between legality and illegality, but different types of informality.

> Such trends point to a complex continuum of legality and illegality, where squatter settlements formed through land invasion and self-help housing can exist alongside upscale informal subdivisions formed through legal ownership and market transactions but in violation of land use regulations. Both forms of housing are informal but embody very different concretizations of legitimacy. The divide here is not between formality and informality but rather a differentiation *within* informality.
>
> (Roy, 2005:149)

Rome's complexity is largely the result of extensive informal practices which express some of the complexity outlined by Roy in her study of India. In Rome most informal housing was built by individual families who lacked the resources required for a proper home. These buildings are termed 'autocostruiti' (self-built) and occupy extensive parts of Rome's territory (Martinelli, 1986:25). Given the informal character of these buildings, the neighborhoods they comprise lack the infrastructure of a functioning city, as well as a coherent street system. These areas are generally subsumed under the term borgata, although their historical formation does not justify the usage. Ferrarotti estimated that in 1970 some 600,000 people lived in illegal housing without any proper urban infrastructure. In addition to these underprivileged residents, another 100,000 citizens lived in the borgate properly speaking (Ferrarotti, 1979:49–50). This constituted approaching 40% of total Roman residents, living in precarious conditions and in need of significant urban interventions to achieve acceptable living standards. The foundation for this assistance was delivered by the 1976 elections.

It was in these circumstances that the O Zones came into being. This was a zoning classification adopted in 1976 to identify areas of informal housing which was not to be demolished, but rather brought gradually into a condition of legality. The process was lengthy, and carries on in the present day. Informal homes can acquire legal status if owners comply with procedures established by a series of laws which have changed over time. The process of identifying and regularizing areas of informal growth started in 1976, and by 1983 74 nuclei of informal growth had been documented with precision.[14] But while informal and illegal growth slowed after 1976 it did not stop. In 1983 another eight nuclei were added to the list, followed by others in 2006.

According to a recent study, of the 12 million residential buildings present in Italy today, only 1.3 million were designed by architects (CRESME, 2017). The rest (41%) were either self-produced by builders lacking professional qualifications, or designed by draftsmen[15] (40%), or by engineers (8%). The minority of buildings designed by architects includes such public projects as Quarticciolo, Trullo, Pamphili and Tiburtino IV. National statistics show that in 2017 some 20% of residential units built that year were illegal (ISTAT, 2018:125). Informality is still strong. Rome stands out as a negative example of this process, often involving graft and corruption. Recently, functionaries working for the municipal agency specialized in regularizing irregularly built real estate were convicted for criminal actions relating to their professional services (Corriere della Sera, 2010). In 2017 a prominent consultant to Rome's mayor was arrested owing to alleged corruption by a real estate developer (Fatto Quotidiano, 2016). Corruption evidently plays a significant role in structuring urban growth.

While in the 1970s the Region of Lazio started to experiment with legislation allowing illegally built housing to be regularized, real progress in the regularization process was only possible with a specific national law of 1984, followed by others in 1994 and 2003. According to their own data, the city of Rome regularizes about 2,000 illegal buildings every month, and rejects only about 100 requests for regularization in the same period, or 5% of the total. City of Rome data show that between 1984 and 1994, 338,943 regularization procedures were approved, 1,629 rejected. In the years following 1994, 170,081 requests for regularization were presented, to which we must add the 84,745 requests for regularization based on the 2003 amnesty law. That makes for a total of almost 600,000 instances of illegal building, almost all of which have been regularized under various laws passed over the last 45 years. Most of this illegal building has taken place in the periphery, especially in the poorer eastern fringes of the city's territory.[16] But illegal building is also found in the more affluent northern districts, and along the coastal areas. Current national statistics show that Rome has 1,370,210 residential buildings (ISTAT, 2011), of which almost half were constructed without appropriate authorization.

Exacerbating this forlorn picture, we must note that a portion of *legal* building was constructed without taking into consideration appropriate standards of urban growth. Corrupt city politics is usually ascribed with responsibility for this trend. The most famous example of this is the Cavallieri Hilton hotel built on Monte Mario, a protected green area to the north of the city center. Monte Mario was occupied in part by Villa Madama, the pristine environment Claude used to capture images of the Roman campagna. The hotel met with significant opposition by activists struggling to protect Rome's surviving green areas. In the end, the builders prevailed, according to some claims because the DC mayor at the time accepted bribes to approve the work (Insolera, 1976:213). A similar case of poor-quality authorized growth is the low-income neighborhood of Magliana. The regeneration plan implemented by the city of Rome in 1993 notes that Magliana was built in 'controversial circumstances' that led to the creation of a neighborhood having the highest density of any part of the city. It has 75,000 inhabitants per square kilometer as compared to the overall city average of 2,000 (Comune di Roma, 2005:2). Like so many parts of the periphery, Magliana started out as a malarial basin later developed for farming. The reclamation process started in 1929, and then encountered difficulties with floods of 1937, and the outbreak of the war in 1940, which led to the conscription of the farmers working the land. These difficulties coincided with news of the fascist plan to organize urban growth out to Ostia following the course of the Tiber (Annappo, 2003). Adversity on the one side and opportunity on the other prompted the owner of Magliana to shift from agriculture to real estate. The possibility of developing this as a site of urban growth was first discussed with city officials in 1953, when it was established that a massive landfill would be required to raise this low-lying floodplain above the level of the Tiber before it could be developed. Some parts of the basin were seven meters below the flood level, and the estimated cost of the landfill was exorbitant. The exact details of the informal arrangements were never documented, but the developer was eventually authorized to build an extremely dense high-rise community without providing the necessary landfill (Annappo, 2003). Today Magliana is well below the flood level, and all that separates residents from the Tiber is a massive levee.

In the periphery we find entire districts of poorly regulated growth. An example is Borghesiana out to the far eastern extremity of the city's territory. Borghesiana started as a PEEP with a core of public housing (Insolera, 1976:287), and was then almost entirely engulfed by informal construction. This neighborhood has one of the highest numbers of requests in the city for the regularization of illegal building.[17] Scattered amongst completed palazzine is a remarkable variety of abandoned lots, informal housing in various states of completion and dusty country roads. This is an example of what Agnew translates as oil stain growth (Agnew, 1995:135). Massimina is even more striking from the standpoint of illegal and unsound growth. On the western fringe of the city outside the GRA ring road, it is hemmed in by

Figure 2.3 Massimina in Via Serafino Belfanti looking west. (Photograph taken in August 2020 © Gregory Smith.).

agricultural land, with an irregular topography extreme by even the local standard (Figs. 2.3 and 2.4). The undulating terrain is cut through by roads so narrow that service vehicles can hardly pass. The roads lack any logic as a network, with a significant number of dead end streets; the buildings express a staggering variety of shapes, sizes and styles (Fig. 2.5). To compound the matter, Massimina is located within half a kilometer of the world's largest landfill, only closed in 2013 and now awaiting regeneration. Yet, notwithstanding these challenges – or perhaps because of them – Massimina has an active neighborhood citizen group, and has attracted the interest of both the local borough and the city planning office.[18] The neighborhood may be poised on the edge of a major progressive transformation.

Micro-quarters

Rome is characterized by such variety as to make generalization impossible. It is an archipelago city, according to a survey of the city commissioned in the early 1990s. This study found that much of daily life in Rome is lived at the level of 198 neighborhoods. Herzfeld notes the importance of this study, stating that these micro-quarters 'in fact correspond closely to the perceptions and experiences of local residents' (Herzfeld, 2009:72). The study helped pave the way to the 2008 master plan; an effort to start from the bottom up to identify a new administrative framework. The micro-quarters were defined using citizen interviews, historical and morphological analysis

Figure 2.4 Massimina in Via Serafino Belfanti looking east. (Photograph taken in August 2020 © Gregory Smith.)

of the urban fabric, the study of local functional requirements and territorial divisions used in the real estate market. Each neighborhood has its own territory, a public image and most have a distinctive identity and name

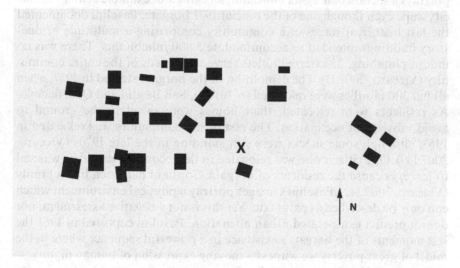

Figure 2.5 Figure ground representation of Massimina. The letter X shows the point from which photographs in Figs. 2.3 and 2.4 were taken. This diagram reveals the parataxic organization so characteristic of Rome's periphery. (©Gregory Smith.)

(Fratini, 2000:30). Taken as a pure arithmetic calculation we can surmise that each micro-quarter has a population of some 14,000 citizens, that of a small town. Most micro-quarters offer a complete range of commercial services, and many municipal ones. This is the level at which everyday life takes place, and the setting for many urban narratives. Many of these neighborhoods are also the object of careful sociological and anthropological studies, themselves evidence of rich concern for local level life in the nation's capital (e.g., Scandurra, 2007).

Variation among these neighborhoods is significant in every regard, including aesthetics. We have noted how the classical figure ground paradigm works well for the center, and many publicly commissioned housing projects. Yet other neighborhoods hardly measure up to this expectation, and must either be rejected as Junkspace, or celebrated as expressions of parataxis. A splendid example of parataxis is found in Pigneto, the chief setting for Pasolini's film *Accattone*. Although once considered marginal and underprivileged, in the last 20 years the neighborhood has attracted public and private investment, and significant gentrification (Scandurra, 2007). Here one finds palazzine, villini, even the occasional shack. Pasolini's historic presence is well publicized, including a bar referred to as Pasolini's bar.[19] The film *Accattone* (Pasolini, 1961) implicitly celebrates the parataxis of the city, especially in the early scenes, where we witness a stunning variety of building shapes, irregularly placed on a disjointed urban terrain.

Another compelling portrayal of parataxis in the same film is a dolly shot which has exercised much film criticism (Rhodes, 2007:50). This sequence portrays a section of borgata Gordiani, an egregious example of an irregular cityscape even though one of the official 1931 borgate. Pasolini documented the last historical stages of a community comprising a multitude of one-story buildings intended to accommodate 5,000 inhabitants. There was no indoor plumbing; 25 external toilets served the needs of the entire community (Viccaro, 2007:11). The demolition of the borgata started in 1957, when all but 300 families were relocated to Tufello, San Basilio and Quarticciolo. As residents were relocated, their homes were razed to the ground to avoid subsequent occupation. The rest of the community was relocated in 1963, although some shacks were still standing in the late 1970s (Viccaro, 2007:139). One citizen who was relocated in the second wave said he was sad to leave, because the residents of borgata Gordiani had been like a family (Viccaro, 2007:142). Pasolini's images portray a physical environment which can only be described as parataxic. Yet this is not a negative assessment, nor does it predict exasperated urban alienation. Pasolini captured in 1961 the last moments of the borgata's existence in a powerful sequence where in the midst of great poverty we witness a moving expression of human dignity.

Tuscolana II constitutes another important Roman neighborhood, centering on an INA housing project designed in the early 1950s by Mario De Renzi and Saverio Muratori. Physically the project adopts a modernist style, boasting a façade with an open reinforced concrete frame filled in with tufa blocks. The project is often used today as a filmic icon of the Roman periphery, and was the chief location for Pasolini's film *Mamma Roma* (Pasolini, 1962). The design is stunning, with a façade which opens as a triumphal arch with arms flung apart to embrace the countryside which once extended from its northeastern flank. The interior of the complex, contained by the building serving as a public façade and locally called the boomerang, has a long irregular central spine, with variously shaped row houses organized in perpendicular fashion, all encircled by tower blocks local citizens call skyscrapers. The variety of architectural forms creates a dramatic and varied urban poché.

The fragments which compose the city are too numerous for us to explore all of them here. Each is part of a changing assemblage bringing together diverse factors, and given life through the actions of ordinary citizens. They are often the settings for important works of art, including the literary and filmic productions of Pier Paolo Pasolini.

Notes

1 Today agro romano and campagna romana are often used interchangeably. For a period starting in the middle ages the ancient term agro was replaced by campagna (Martirano and Medici, 1988:449). In modern official designations the most peripheral areas of the municipal territory are now classed as the agro romano, of which there are 59 zones (Comune di Roma Ordinance of 13 September 1961). According to the classification documented by Lugli, these peripheral zones should be properly termed campagna romana.

2 Legge 491 di 17 luglio 1910 Provvedimenti per estendere il bonificamento e la colonizzazione nell'agro romano.

3 http://www.archiviocapitolino.it/files/archivio/ripartizione_v_lavori_pubblici_-_ufficio_agro_romano.pdf (Accessed June 12, 2019).

4 Up until a few years ago the façade at Via Fonteiana 86 bore a plaque commemorating Pasolini's stay. By 2019 the plaque had been removed, leaving only the holes once used to secure it. At Via Carini 45 one still finds a commemorative plaque. Evidently Pasolini's memory is still controversial.

5 Today some of the apartments are being sold on the private market. I had occasion to interview a young couple who bought one unit, and were delighted to be part of a housing project which played such an important role in Italian literature.

6 The area where Quarticciolo was built had previously been farmland, and at 7.6 kilometers from the ancient Golden Milestone was fully within the area of the ancient Roman agro. The land had belonged from remote times to the patriarchal basilica of Saint Mary Major, which donated the property to the city of Rome, some 390 hectares, in 1939 (Nocera, 2010). The availability of an extensive building site free of cost is what determined its location.

7 I spent four months in spring 2013 exploring Quarticciolo with a group of Cornell University undergraduate planning students.

8 http://www.teatriincomune.roma.it/teatro-biblioteca-quarticciolo/la-storia/ (Accessed November 11, 2019).

9 The floodplain on the western side of the Janiculum Hill is clearly visible in a map published by Heiken et al. (2007:134).

10 http://www.urbanistica.comune.roma.it/pdz/elenco.html (Accessed November 11, 2019).
11 The requalification project can be consulted at this URL: https://www.lauraperettiarchitects.com/project/rigenerare-corviale/ (Accessed November 11, 2019).
12 https://geoportale.regione.lazio.it/geoportale/ (Accessed November 11, 2019).
13 I spent spring 2019 exploring Bravatta Pisana with my Cornell University planning students.
14 http://www.urbanistica.comune.roma.it/zoneo.html (Accessed November 11, 2019).
15 This refers to a geometra, a professional status achieved by completing studies at a specialized high school.
16 http://www.urbanistica.comune.roma.it/zoneo.html (Accessed November 11, 2019).
17 http://www.urbanistica.comune.roma.it/zoneo.html (Accessed November 11, 2019).
18 The neighborhood group has a website at http://cdqmassimina.it/ (Accessed November 11, 2019).
19 https://ricerca.repubblica.it/repubblica/archivio/repubblica/2009/04/01/roma-brucia-il-bar-amato-da-pasolini.html (Accessed November 11, 2019).

Bibliography

Accasto, Gianni, Vanna Fraticelli and Renato Nicolini (1971) *L' architettura di Roma capitale: 1870-1970*. Rome: Golem.

Agnew, John A. (1995) *Rome*. Oxford: John Wiley.

Annappo, Antonello (2003) *La Tenuta Pian Due Torri*. Rome: Fondo Riva Portuense.

Banfield, Edward C. (1958) *The moral basis of a backward society*. New York: Free Press.

Beaven, Lisa (2014) 'The grave of graves. The response of grand tourists to the Roman campagna,' in Bonifas, Gilbert and Martine Monacelli (eds.) *Southern horrors: northern visions of the Mediterranean world*, pp. 79–93. Newcastle upon Tyne: Cambridge Scholars Publishing.

Berruti, Gilda (2019) *Fuori norma. Percorsi e ragionamenti su urbanistica e informale*. Rome: INU.

Bobbio, Norberto (2006) *Democracy and dictatorship. The nature and limits of state power*. Translated by Peter Kennealy. Cambridge, UK: Polity Press.

Carocci, Sandro and Marco Vendittelli (2004) *L'origine della campagna Romana. Casali, castelli e villaggi nel XII e XIII secolo*. Rome: Società Romana di Storia Patria.

Casalini, Simona (2017) 'Muore Teodoro Buontempo. La destra romana in lutto,' *La Repubblica*, 27 February.

Cavallo, Federica Letizia (2011) *Terre acque macchine. Geografie della bonifica in Italia tra ottocento e novecento*. Reggio Emilia: Diabasis.

Celli, Angelo (1984) 'Contadini nella campagna romana,' in Bevilacqua, Pietro and Manlio Rossi-Doria (eds.) *Le bonifiche in Italia dal '700 a oggi*, pp. 234–266. Bari: Laterza.

Comune di Roma (2005) *Programma di recupero urbano Magliana*. Rome: Comune di Roma.

Corriere della Sera (2010) 'Bancarotta fraudolenta: ecco dove finivano i soldi della Gemma, S.p.a.,' *Corriere della Sera*, 23 November.

CRESME (2017) *Chi ha progettato l'Italia? Ruolo dell'architettura nella qualità del paesaggio edilizio italiano*. Rome: CRESME.

Dal Mas, Roberta (2006) 'Una ricostruzione del processo di formazione del quartiere Monteverde,' in Cassetti, Roberto and Gianfranco Spagnesi (eds.) *Roma contemporanea. Storia e progetto*, pp. 187–196. Rome: Gangemi Editore.

Eramo, Nella (2008) *Mutui per la bonifica dell'agro romano e pontino (1905-1975)*. Rome: Minister for Archives.

Fatto Quotidiano, Il (2016) 'Roma: Marra arrestato,' *Il Fatto Quotidiano*, 16 December.

Felisini, Daniela (2009) 'Viaggio in Italia. L'economia di Roma e del suo territorio nelle descrizioni dei viaggiatori stranieri dell'Ottocento,' in Formica, Marina (ed.) *Roma e la Campagna Romana nel Grand Tour. Atti del convegno interdisciplinare a Monte Porzio Catone Roma 17-18 maggio 2008*, pp. 295–312. Bari: Laterza.

Ferrarotti, Franco (1979 [1970]) *Roma da capitale a periferia*. Bari: Laterza.

Fratini, Fabiola (2000) *Roma arcipelago di isole urbane. Uno scenario per il xxi secolo*. Rome: Gangemi Editore.

Freidrichs, Chad (dir.) (2011) *The Pruitt-Igoe myth*. New York: First Run Features.

Gisotti, Maria Rita (2008) *L'invenzione del paesaggio toscano. Immagine culturale e realtà fisica*. Florence: Edizioni Polistampa.

Guccione, Margherita, Maria Margarita Segarra Lagunes and Rosalia Vittorini (2002) *Guida ai quartieri romani INA CASA*. Rome: Gangemi Editore.

Heiken, Grant, Renato Funiciello and Donatella De Rita (2007) *The seven hills of Rome. A geological tour of the eternal city*. Princeton: Princeton University Press.

Herzfeld, Michael (2009) *Evicted from eternity: the restructuring of modern Rome*. Chicago: University of Chicago Press.

Insolera, Italo (1976) *Roma moderna. Un secolo di storia urbanistica 1870-1970*. Bari: Laterza.

ISTAT (2011) *15° Censimento della popolazione e delle abitazioni 2011*. Rome: ISTAT.

ISTAT (2018) *Rapporto BES: il benessere equo e sostenibile in Italia*. Rome: ISTAT.

Jencks, Charles (1977) *The language of post-modern architecture*. New York: Rizzoli.

Krautheimer, Richard (1980) *Rome: the profile of a city 312 - 1308*. Princeton: Princeton University Press.

Lizzani, Carlo (dir.) (1960) *Il Gobbo*. Rome: Dino de Laurentis.

Lugli, Piero Maria (2006) *L'Agro Romano e l'"Altera Forma" di Roma antica*. Rome: Gangemi.

Martinelli, Franco (1964) *Ricerche sulla struttura sociale della popolazione di Roma (1871-1861)*. Pisa: Libreria Goliardica.

Martinelli, Franco (1986) *Roma nuova. Borgate spontanee e insediamenti pubblici. Dalla marginalità alla domandi dei servizi*. Milan: Franco Angeli.

Martini, Mauro and Anna Parasacchi (2004) *Intervista a Corviale. L'esperienza di un laboratorio per lo sviluppo locale e la partecipazione*. Rome: Comune di Roma.

Martirano, Giovanni and Riccardo Filiberto Medici (1988) 'L'Agro romano: un intenso popolamento,' in Barberis, Corrado and Gian Giacomo Dell'Angelo (eds.) *Italia rurale*, pp. 290–309. Bari: Laterza.

Muntoni, Alessandra (2006) 'La Roma delle 'palazzine' dagli anni venti agli anni sessanta,' in Cassetti, Roberto and Gianfranco Spagnesi (eds.) *Roma contemporanea. Storia e progetto*, pp. 187–196. Rome: Gangemi Editore.

Mura, Maria Luisa (2014) *Testaccio. Il XX Rione di Roma*. Rome: Edilazio.

Nicolini, Renato (ed.) (2010) *Roberto Nicolini architetto 1907-1977*. Rome: Prospettive.

Nicolini, Renato (2011) *Estate Romana: un effimero lungo nove anni.* Reggio Calabria: Città del Sole Edizioni.

Nocera, Alessandro (2010) 'Progettare borgate. Il Quarticciolo,' in Nicolini, Renato (ed.) *Roberto Nicolini Architetto 1907-1977*, pp. 35–38. Roma: Prospettive.

Pasolini, Pier Paolo (dir.) (1961) *Accattone.* Rome: Cino Del Duca.

Pasolini, Pier Paolo (dir.) (1962) *Mamma Roma.* Rome: Arco Film.

Passigli, Susanna (2012) 'La costruzione del "Catasto Alessandrino" (1660). Agrimensori, geometri, periti misuratori,' in Bevilacqua, Mario and Marcello Fagiolo (eds.) *Piante di Roma dal rinascimento ai catasti. Atti del convegno internazionale, Roma, novembre 2010*, pp. 370–391. Rome: Editoriale Artemide.

Regni, Bruno and Marina Sennato (eds.) (1982) *Innocenzo Sabbatini. Architetture per la città.* Rome: Architettura Arte Moderna Edizioni.

Renga, Dana (2017) 'Remediating the Banda della Magliana: debating sympathetic perpetrators in the digital age,' in Pickering-Jazzi, Robin (ed.) *The Italian antimafia, new media, and the culture of legality*, pp. 137–161. Toronto: University of Toronto Press.

Rhodes, John David (2007) *Stupendous miserable city. Pasolini's Rome.* Minneapolis: University of Minnesota Press.

Roy, Ananya (2005) 'Urban informality: toward an epistemology of planning,' *Journal of the American Planning Association*, 71:2:147–158.

Sanfilippo, Mario (1993) *Le tre città di Roma. Lo sviluppo urbano dalle origini a oggi.* Bari: Editori Laterza.

Scandurra, Giuseppe (2007) *Il Pigneto. Un'etnografia fuori le mura di Roma: le storie, le voci e le rappresentazioni dei suoi abitanti.* Padua: Cleup.

Sereni, Emilio (1997) *History of the Italian agricultural landscape.* Translated by R. Burr Litchfield. Princeton: Princeton University Press.

Stabile, Francesca Romana (2012) *La Garbatella a Roma. architettura e regionalismo.* Rome: Dedalo Edizioni.

Strappa, Giuseppe (ed.) (1989) *Tradizione e innovazione nell'architettura di Roma capitale 1870-1930.* Rome: Edizioni Kappa.

Turri, Eugenio (2014) *Semiologia del paesaggio italiano.* Venice: Marsilio.

Valenti, Ghino (1984) 'Il sistema agrario dell'agro romano. Leggi speciali,' in Bevilacqua, Paolo and Manlio Rossi-Doria (eds.) *Le bonifiche in Italia dal 700 ad oggi*, pp. 212–233. Bari: Laterza.

Venditti, Emilio (1989) *Il Trullo 1939-1989.* Rome: Tipografia Trullo.

Viccaro, Ulrike (2007) *Storia di borgata Gordiani: dal fascismo agli anni del "boom".* Rome: Franco Angeli.

Villani, Luciano (2012) *Le borgate del fascismo. Storia urbana, politica e sociale della periferia romana.* Milan: Ledizioni.

Wine, Humphrey (1994) *Claude: the poetic landscape.* London: National Gallery Publications.

Zannoni, Osvaldo (2010) *Il tramsporto del tranviere.* Cortona: Calosci.

Zeppegno, Luciano (1996) *I rioni di Roma.* Rome: Newton Compton Editori.

3   Pasolini's Rome

Introduction

In this chapter I have two aims. One is to illustrate the method Pasolini employs to pursue his ethnographic explorations of the Roman periphery in the 1950s. The other is to describe his characterization of life in these same areas. A salient feature of his characterization is persisting tension between a principle of civility on the one side, and violent antisocial behavior on the other. Permeating this tension is a quality I define as the sacred, implicitly referenced throughout his portrayals. This is life beyond any effective system of control, be it the state or the party or the church, or any form of economic integration. These citizens are left to their own devices in adverse physical conditions worsened by a moral system which in no way eases their plight. Yet against all odds they still preserve a deep sense of human dignity. The painstaking detail found in these accounts sacralizes the habitus in which social life unfolds, and invites the reader to take seriously Pasolini's claim to describe life as it is actually lived. These settings are so remote from the experience of his chiefly middle-class readers, that few would have ever heard them mentioned before. It is uncharted territory, exposed in a bold and courageous expedition.

To the modern reader the most shocking feature of Pasolini's books is the extreme gendered violence and profound family pathology.[1] These portrayals might appear to be a form of negative idealization, a critique of the moral system promoted especially by the Catholic Church. Yet if we draw comparisons with ethnographic research conducted among poor families in Naples in the same period, we see that the poet's findings reflect actual conditions of family life also documented in another part of the Italian south. I am referring to the research of Anne Parsons, an exceptionally gifted and well-qualified academic researcher. She provides a vision of family and community life in Naples which is strikingly similar to that furnished by Pasolini. The violence revealed in Rome can be traced to a set of antisocial norms also found in distant urban settings characterized by significant marginality, such as the American inner-city ghetto. A modern Marxist explanation for such conduct resonates with Pasolini's implicit analysis of the 1950s.

Pasolini wrote for a broad public, and while wishing faithfully to describe the life he had experienced in peripheral Rome, he takes minor poetic liberties. Here I explore in particular liberties taken with spatial relationships used to build narrative tension. Finally, we must note that while Pasolini treats the periphery as a homogeneous area of exclusion, some of the episodes he describes reveal that social and political marginality coexists in close proximity with privilege and inclusion. Thus, while focusing on marginality, Pasolini provides witness to the complex geography of the city that mixes together widely different social groups.

Pasolini as ethnographer

I use the term ethnographer in a broad sense. Ethnography is part of an important European current in the social sciences, especially associated with the United Kingdom and the United States. Naturally, intellectual boundaries have abated, and a mix of disciplinary approaches is today found in any national setting. But in the 1940s and 1950s, when Pasolini began his field explorations, national context had weight. The 1950s was a period of dramatic change for Italy, witnessing the precipitous rise of a neourban society to the detriment of traditions which had locked Italians for centuries into their distinctive local territories. This transformation prompted a host of efforts to understand the emerging character of the nation. Fascination with the multiplicity of Italy's cultures has deep roots in Italian realist literature, the 'verismo' (realism) with which Pasolini can be associated (Asor Rosa, 1979). Typically, these literary investigations were concerned with rural settings, and Pasolini's innovation was to seek diversity at a closer range. Other authors, like Carlo Emilio Gadda (especially Gadda, 1957 [1946]), had also explored the social and cultural extremes of a modern urban setting. But Pasolini, with his systematic investigations, revealed a unique concern for the problem of objectivity, articulated within an overarching theory of society. His theoretical findings have had cogent and lasting explanatory power.

The most noted academic field researcher in Italy at that time was Ernesto de Martino, whose studies of religious cults in the Italian south gained currency starting in the late 1940s. Other field explorations were produced in those years, including the ethno-musicological research of Diego Carpitella, and explorations carried out by the American folklorist Alan Lomax. An important forum encouraging exploration of the country seen through the eyes of its people was the Italian Ethnological Center, founded in 1954 by Ernesto de Martino, Diego Carpitella and Franco Cagnetta (Cagnetta, 2002:19). De Martino was a noted ethnographer, while Carpitella was an ethnomusicologist. Cagnetta was instead an eclectic figure, who moved between Italy, France and the United Kingdom, publishing not only ethnography, but also using his findings in artistic events (Cagnetta, 2002). In 1954 Cagnetta published his study of a remote community in Sardinia in *Nuovi Argomenti*, a journal co-directed by Alberto Moravia, to which

Pasolini also contributed. Not incidentally, Cagnetta carried out an investigation in the early 1950s of prostitution in the Roman slum of Mandrione (Cagnetta, 2002:20). Prostitution was then practiced in brothels made legal with the unification of Italy, and illegally on city streets. Prostitution was widely practiced, but hidden by a veil of Catholic morality scrutinized in those years in numerous investigations, including those of Senator Lina Merlin, whose activity led to the suppression of brothels in 1958 (Merlin and Barberis, 1955). Pasolini became something of an expert in these hidden aspects of marginal life, and was invited to consult on various films portraying prostitution, most famously Federico Fellini's *Le notti di Cabiria* and Carlo Lizzani's *Il Gobbo*. Pasolini also offers compelling portrayals of prostitution in his novels and films.

In those years Ernesto de Martino provided a model for investigating the marginal life of the nation. He was a complex intellectual with a Crocean past, Marxist interests and a conflictual relationship with the Italian Communist Party.[2] His work on southern Italy was much read in his day, and Pasolini collected his published work already in the early 1950s (Chiarcossi and Zabagli, 2017). De Martino brought to public attention surprising – and even shocking – practices which revealed just how distant many remote southern communities were from the life of the more 'advanced' parts of the nation. Researchers like Ernesto de Martino were not only detached observers, and in addition to documenting these conditions of apparent backwardness, they offered reflections on how things might change. De Martino expressed the hope that integration in a national consumer economy would rescue poor peasants from their plight.[3] Although in the end his historical analysis proved accurate, this was precisely the opposite of what Pasolini hoped for when discussing both the urban underclass and the rural south.

Pasolini's interest in documenting life in marginal Italian communities precedes his arrival in Rome. His first novel, written in the 1940s but only published years later, is a narrative study of peasant culture in Friuli. The book's title is *Il sogno di una cosa* (*The dream of a thing*) (Pasolini, 1962a), a citation taken from Karl Marx where the object of the dream is revolution (Bravo, 1965). It is in this period that Pasolini begins to perfect his skill in documenting life in marginal communities. An understanding of these conditions is exposed in particular by recording the language used by living subjects in real settings. These details are then combined to form a narrative. The Friulian novel is indeed remarkable for its exact detail and powerful evocative prose.

The Einaudi edition of *Street kids* (Pasolini, 1979) provides two useful appendices which explain Pasolini's research methodology, stating that the author had spent between 1950 and 1955 'fully immersed' in the culture of the borgata. One of the appendices, published in 1951 during Pasolini's period in Friuli, provides special attention to the role of language in his investigations. This appendix is entitled *I parlanti* (*The speaking subjects*).

The other appendix, *Il metodo di lavoro* (*The method of investigation*), first published in 1958, is concerned with the Roman investigations.

The method opens with reference to Leo Spitzer. Spitzer had been a student of Ferdinand De Saussure, whose intellectualism he rejected and focused instead on language's affective character (Schiaffini, 1954:1). Pasolini discusses the similarities between his two Roman novels – one already published and the other due to be released the following year – explaining that they were both drawn from the same set of experiences inspired by a uniform investigative methodology. He had intended to write a trilogy conceived before he published the first volume which aimed to serve as an overture to the three. But only two of these novels were completed before Pasolini turned his attention to film. Spitzer had a cult following in Italy in the 1950s, and would have been well known to sophisticated readers (Wellek, 1960:311). But since his only publication in Italian came out in 1954 – attracting public acclaim the following year – Spitzer's work could hardly have inspired the actual methodology for research Pasolini started in 1950. *The method* appears to be chiefly an ipso facto reflection, although it may have guided Pasolini's final editing of *Street kids* which came out in the spring of 1955.

Having won the 1955 International Feltrinelli Award from the Lincei Academy, the country's most prestigious linguistic institution, Spitzer provided authoritative support for an innovative approach to documenting life experiences as a text which could be rendered in narrative form (Wellek, 1960:311). Spitzer was interested in the psychological root of literature, and its ability to document the author's soul. Spitzer rejected the idea of trying to grasp the overall picture of the analyzed text, and instead started his explorations by documenting single details. This circular process (the philological circle) consists in passing from specific details to the formulation of a general hypothesis concerning how the literary piece represents the soul of the analyzed text (Wellek, 1960:315). Adapted to a community study, this is precisely the method used in Pasolini's first novel, *Street kids*, which brings together various observations with no apparent plot or analysis. It is a pure reading of life as text, represented in a way that lays bare the soul of young boys living in the Roman borgate. When the book came out, the prominent literary critic Carlo Bo (1955) claimed that the book contained the flaws of any writer's first novel. Possibly, rather than flaws, this was the consequence of a deliberate effort to implement Spitzer's philological circle in the study of marginal social life.

The method not only describes his approach to documenting the periphery, it also says something about the subjects of the study. These are ordinary young boys,[4] abandoned to themselves, living outside the world controlled by middle-class norms. They are marginal, and in their marginality continue the life of a prehistoric world detached from modernizing trends. Pasolini explores this idea of a surviving prehistoric age throughout his writings, as a powerful critique of the massive changes experienced in the 1960s. On a par with peasants, Pasolini believes, the Roman underclass is imbued in an oral tradition reaching back to the remote regions of European history.

The method stresses a full-immersion research approach – a form of participant observation – rather than interviews which would create impersonal subjects forced to perform on demand. His approach penetrates into the life of the periphery as actually lived. He mentions his deep familiarity with disparate peripheral neighborhoods, among which Torpignattara, Alessandrina, Torre Maura, Pietralata. His recording technique consists of jotting down notes, especially on linguistic expressions. These are vernacular idioms recorded directly on the speaker's lips in the living circumstances of their deployment. When he needed help in understanding local dialect, he turned to his friend Sergio Citti, a house painter from the periphery whom he met in 1951.[5]

As in Fruili, Pasolini wishes to preserve and celebrate a minor though important linguistic heritage, an effort highlighted in appendices to both Roman novels furnishing translations from dialect to Italian. He acknowledges that he views life in the periphery from a bourgeois standpoint, crossing a profound cultural barrier in the process. He immerses himself in the personality of his subjects to capture precisely the true spirit of people he exposes to careful sociological and psychological scrutiny. This is not a cynical undertaking, nor speculation on the suffering of the poor, he specifies. Nor is it the work of a dilettante. In this connection he mentions that he avoids following his 'eros' in order not to traumatize bourgeois sensitivities and jeopardize the success of his research.[6]

His research also has a personal imperative, namely the necessity to come into contact with pristine humanity. He specifies that this personal imperative does not, however, inhibit him from generating documentary data 'in an objective way.' These data are the foundation for his writings, and witness to conditions of marginal life. His research has two aims. One is naturalistic-documentary: to generate objective knowledge of these communities. The other is sensual-stylistic: to collect material for his artistic representations.

In characterizing the subject of his study, he states that from a rational standpoint the underclass is inferior to the bourgeoisie. Here he is referencing ideas expressed by other scholars writing about the south, like Ernesto de Martino. Yet Pasolini immediately adds that in terms of irrationality – or pure vitality – the underclass is superior. This statement neatly sums up his effort to create a Marxism of the irrational, and also justifies his need to experience the pure vitality of the peripheral world. In a characteristic jibe at the PCI, he tells us that his work is different from that of a party leader who acquires knowledge as a premise for transformative political action. Giovanni Berlinguer (1955) was an example of such a writer, a high-ranking representative of the PCI who wrote about the Roman periphery. For Pasolini this research was manipulative, while his own investigations were concerned to celebrate the deep human vitality to which the PCI was blind. The pursuit of liberty and justice does not always, Pasolini claims, lead to the happiness of moral plenitude. Here he references the dilemma he later theorized in greater detail about how to help the poor without destroying their timeless qualities.

Pasolini's letters also provide insights into his research method. In particular we have a letter written in the summer of 1952 to one of his closest friends, Silvana Mauri (Pasolini, 1986:486). The letter discloses something about Pasolini's research technique, as well as deep empathy for the subjects of his study. Take the case of a young boy he met in Villa Borghese (Fig. 3.1). The boy's name was Cristoforo, and his father had died during the war, followed by the death of his mother in 1948. Cristoforo was thus left homeless and destitute. Odd jobs allowed him to get by, and when he had money, he could take a room in a private home. But for some eight months of the year he slept in the open air, either in Villa Borghese or under one of the Tiber bridges. Pasolini was concerned about the boy, and asked Silvana for advice in seeking placement in a child shelter north of Rome (Villaggio del Fanciullo). The letter indicates that he had already referred another child to this facility. In a later letter, we learn that Pasolini had lost contact with Cristoforo. His efforts to track down the boy by talking to other park residents were thwarted by their fear that he might be trying to reach the

Figure 3.1 Villa Borghese with a view of Saint Peter's Basilica. (Photograph taken in September 2020 © Gregory Smith.)

boy owing to theft or some other criminal activity (Pasolini, 1986:489). In the event, having lost contact, Pasolini was unable to move forward with assistance.

These letters not only reveal empathy, but also recount episodes which find their way into his novels. In the first letter to Silvana, he describes an episode conveyed by Cristoforo where someone had stolen the shoes right off his feet as he slept in the open air. In *Street kids* the protagonist sleeps on a bench in Villa Borghese, and awakens to find that his shoes had been stolen. This is the episode described in the novel.

"Hey, did I take off my shoes last night?" Riccetto asked in a loud voice, sitting up suddenly. "No, I didn't take them off," he answered, looking under the bench, on the grass, among the bushes. "Caciotta, Caciotta," he yelled, shaking Caciotta, who was still sleeping, "they stole my shoes!"

(Pasolini, 2007:72)

In the same letter, Pasolini describes how Cristoforo, though not a thief at heart, was compelled to live by expedients, including theft. Cristoforo stole a suitcase from a Swiss pilgrim, and lived from the proceeds of the theft for six months. This is the source for an episode told in *A violent life* where street kids steal a suitcase from a car belonging to a German pilgrim, and fence the contents in Trastevere before carrying on with their nocturnal adventures.

Different from the other boys, Cristoforo is described as being well-bred ('educato'). He had internalized the image of his mother, and with it the memory of a 'normal' life. This normal background provides moral comfort, but also inhibits his ability to become scum like the other boys. By an irony of fate, Cristoforo was destitute, but bereft of the carefree spirit which made real street kids happy even when they had not eaten for days. Real street kids are supported by an unexplainable happiness that causes them unexpectedly to break out in song. This apparent carefree spirit is also a response to tension between an inhumane environment and deep intrinsic humanity. The tension is especially evident in the final scene in *Street kids*, when Riccetto witnesses the drowning of a young companion, but keeps on walking, restraining his emotions. Pasolini tells Silvana that these boys live in a 'frightful state' of deprivation and neglect, yet still preserve their peculiar dignity.

Background on *Street kids* and *A violent life*

When reading *Street kids* today, especially in the elegant English-language translation by Ann Goldstein, it is hard to understand how the book could have been seized by magistrates owing to obscenity of language. Yet in 1955, language and sexuality were tightly controlled not only by the church, but also by the Italian Communist Party. The DC had no sympathy for a declared Marxist homosexual who revealed the seamy side of life in Rome's poor districts. Nor, on the other hand, did the PCI wish to be displaced by the DC in safeguarding the morality of the urban proletariat. Giovanni Berlinguer (1955) claimed at the time that *Street kids* was a trivial exercise in aesthetics, a decadent literary product with no bearing on the reality of the periphery. Tonelli asserts that the PCI had a messianic vision of its role in representing the interests of the periphery, and did not tolerate the intrusion of an outsider (Tonelli, 2015:105). In the event, the Presidency of the Council of Ministers in July 1955, supported by a coalition of reactionary parties led by the DC, transmitted a request to the magistrates urging that the book be seized and the author and publisher be incriminated (Tonelli, 2015:108).

Pasolini's vision was incompatible with the image of Italy then being transformed by the 'economic miracle,' that modernizing process which would pave the way to a society of high mass consumption. It was not the obscene language which prompted judicial action, but the threat of a dest-abilizing force which could reveal the country's backwardness and poverty in areas which Enzo Siciliano called 'an existential suburb in Italian soci-ety' (Siciliano, 1978:210). In July public hearings were opened, and evidence marshalled for prosecution and defense. The poet Giuseppe Ungaretti, a Pasolini supporter, testified that the book did no more than represent the real language of peripheral youngsters. Pasolini did well to reproduce lan-guage in the way it was truly spoken, representing in realistic terms the plight of poor citizens, rendering their conditions of existence in high poetic form (Tonelli, 2015:110). Although not called upon to testify, approval for the book also came from Carlo Bo, the distinguished literary critic and future Christian Democrat senator. Not only did the book give a voice to the poor, he claimed, it also expressed a religious quality which drew the poor to their own form of piety. In his review of *Street kids* in the widely circulating weekly *L'Europeo*, Bo (1955) first defines Pasolini as one of the few poets of the current generation with strong character and solid literary foundations. He then describes the book as the most politically committed of Pasolini's 'brilliant career.' He terms the novel 'meditated, justified and valid.' He praises the sophisticated use of language, and the revelation of an unexpected sense of piety in a marginal community. Pasolini was no complacent or complicitous observer, instead he documents in a 'cold and unrelenting way' the braggadocio of desolate citizens who live in a 'shame-ful condition of diminished humanity.'

In the end, Pasolini and the publisher were acquitted of charges and the book was released. The Catholic newspaper, the *Osservatore Romano*, noted with disappointment that the magistrates were no longer able to guarantee common decency. This is the motivation for the sentence.

> The vulgar and trivial words, typical of the periphery, though repeat-edly pronounced, are justified considering the psychology of these young people, and the instincts that drive them. This is in addition to

the character of their desires (and vulgar language is not always the same as obscene language) [...] nor does the author dwell with narcissistic malice on the objectively obscene situations [....].

(Tonelli, 2015:111)

By the time *A violent life* was published in 1959, Pasolini was an established public figure, and his new novel met with none of the outraged response of the first, although the right responded negatively to its publication (Tonelli, 2015:117). The second novel is more analytical in its construction, with a plot especially concerned to discredit the PCI. In one of the final scenes of the novel, the hero, Tommaso, takes out a PCI card, only to discover that the local cell leaders, far from supporting the general interests of the community, are really feathering their own nests.

The first two films on Rome follow the same pattern as his two novels, where *Accattone* (Pasolini, 1961) celebrates in an aesthetic way the sacred qualities of the Roman underclass, and *Mamma Roma* (Pasolini, 1962b) provides a more analytical view, especially in critiquing the reformist aims of the Italian state. His third film, *La ricotta* (Pasolini, 1963), provides a more explicit critique of the contemporary political and economic system, with a suggested indictment of the PCI. His last Roman film, *Uccellacci e uccellini* (Pasolini, 1966), provides a further critique of the PCI.

The angry polemic with the PCI lasted his entire life. More important for posterity, however, is the persisting impact of his vision of the periphery. This is life striving for a raw expression of piety, a moral quality aligned with a pure expression of human vitality. This condition of sacrality is associated with the civic culture which motivates much of everyday life in even difficult conditions of existence.

Civility and incivility

The role of civility in structuring everyday Italian life has been noted by a host of ethnographers. Sydel Silverman (1975) in particular shows the far-reaching relevance of this concept in the life of a small Umbrian town in the early 1960s. That was the period of historical transition between traditional Italy and rapidly emerging modernity. At its core, the sense of civility revolves around the capacity to express empathetic concern for a broader community, and in this interpretation is often rendered in contemporary theory as social capital. Social capital was only theorized in the 1990s, well beyond the times of Pasolini, and indeed Silverman. But both of the latter writers describe civility understood in this way. Geographical variation in the expression of civic commitment has been the topic of considerable debate in Italy. The most famous contemporary model for explaining the uneven distribution of social capital, and identifying the consequences of its discontinuous presence, is certainly that formulated by Robert Putnam

(1993). For Putnam social capital understood as a civic ideal explains all areas of community performance, from politics to economics, and even human happiness. He claims that some parts of the Italian community – notably the Italian south – are burdened by an irreversible lack of social capital.

Putnam's model is of immediate importance to our retrospective reading of Pasolini, since the greater part of the Roman periphery is made up of southerners. Most migrants who moved to Rome's periphery in the 1940s and 1950s came from southern Italian regions (Ferrarotti, 1979:27). To take one case, we can note that according to some estimates 600,000 Calabrians today reside in Rome,[7] making it the largest Calabrian city in the world. Ninetto Davoli, Pasolini's companion for many years, was from a Calabrian family which had moved to the Roman periphery in the late 1940s. Seen from the perspective of Italy's common anti-southern bias, we might expect the periphery to experience a deficit of civic sensitivity. As we have seen, Ferrarotti (1979) follows this reasoning in characterizing peripheral neighborhoods as lacking community spirit. Yet he hints at a historical explanation, suggesting that their civic sense had been undermined in a political strategy aiming to diminish capacity to express a community which could mobilize support for a progressive political project. Pasolini instead identifies an entirely new foundation for civic sensitivity in the periphery, corrupted by bleak conditions of existence, what Bo calls the 'shameful condition of diminished humanity.' Pasolini imagines a condition of pure humanity with compelling community potential, and a unique type of civility blunted in its expression by imposed adversity.

Civility in its classical conception is a shared project historically associated with the city. Pasolini gives many instances of citizen awareness of the sharp boundary dividing townsmen from countrymen in the Roman periphery. This is even evident among recent migrants, in large measure poor farmers who moved recently to the city. Tommaso's life history is described in *A violent life*, showing that his family had come to Rome from the south as war refugees in 1943. Combined Allied and Nazi military efforts had destroyed his family's modest farmhouse, and flight to Rome was their best option. Once established in the city, Tommaso expresses contempt for the countrymen frequently seen on the urban fringe. They are repeatedly derided as 'hicks,' known throughout Italy as cafoni, and in Rome especially as burini or 'butters.' Hicks belong to the countryside, contrasted with the city as a privileged place. Tommaso's identity as an urbanite is expressed in his access to the entire city, from the periphery to the center. An example of fascination for the city is provided in *A violent life* in a scene where Pasolini uses the narrative expedient of a stolen automobile to allow the protagonists to explore the countryside. The narrative takes the travelers to the north of Rome, describing the visual impact of the city as seen from afar and its contrast with the surrounding farmland.

Cutting across the Janiculum and Monte Mario, they were soon in open country, all hilly. After a few miles through meadows and woods, with pieces of Rome shining here and there in the distance [they stopped] with dogs barking all around from the farms.

(Pasolini, 2007:56–57)

A violent life provides witness to numerous encounters between poor peripheral urbanites and farmers. Only rarely do these images harken to the historic bucolic vision, such as the scene showing grazing sheep in the film *La ricotta*. Most representations instead suggest a negative reading of farmers and farm life. Italy, of course, has an antirural bias with illustrious literary roots. Montanari (2009) shows how townsfolk have historically existed in close proximity to the rural poor whom they kept in a subservient position through negative stereotyping. In my own ethnographic work in Abruzzo (Smith, 2013), I note strong hostility between peripheral Romans and the inhabitants of small towns, held up as hicks by transiting marginal urbanites. Running down rural folk is a way to stress urban status. In a similar strategy, Tommaso loses no opportunity to express contempt for farmers. This is evident in the scene where a gas station worker is robbed. His contemptable condition as a rural worker is rendered visible in the narrative, egging on the malefactors in their criminal actions.

[The gas station attendant] must have been a peasant who had come to Rome not long ago, from Abruzzo or from Puglia; you could tell by his broad, sun-baked face, by his mouth that had a dumb expression even in sleep, and also by the strength you could sense within the folds of the unbuttoned overall.

(Pasolini, 2007:60)

Tommaso, though poor, feels his superiority as a town dweller. There is a refinement in the urban condition, even when lived by people in circumstances of exclusion. In film the best expression of this stark contrast is in *Mamma Roma*, where in the opening scene a farmer during a wedding ceremony lifts his glass in a toast delivered in strong country dialect. The terminology he adopts is calibrated to appear ridiculous as seen from the standpoint of even a marginal urbanite. His toast is interrupted by the derisive comments of an urban pimp.

THE BRIDE'S FATHER (IN A STUPID WAY): I want to tell you dear youngsters
    that …. on this day …
THE PIMP: …. out on parole ….
THE BRIDE'S FATHER: … that on this day that we have all come together, I
    want to say ….
THE PIMP: …. That you are an informer! …
THE BRIDE'S FATHER: … that I like all this company. Because even if we work
    the land, we are people … (he strikes his chest sharply several times) …

THE PIMP: Who suffer from tuberculosis!

THE BRIDE'S FATHER: (continues to strike his chest) ...we are people that have a heart ....and (he utters incoherent references to the bible in country dialect, without any logic to them) ...

(Pasolini, 2006:242–243)

The dwellers of the periphery may be deadbeats ('morti di fame'), as we hear them say of themselves repeatedly in *A violent life*, yet they are still city dwellers, and as such obey the rules of civil conduct. Women in *Street kids* are poor yet make an effort to show their concern with decent appearance, especially in the presence of boys.

> [....] at that moment her sister came in, all dolled up, the one who was eighteen. She had taken a while to enter because she had put on her good dress, of black silk, and had even added a little lipstick.

(Pasolini, 2016:131)

Men also express concern for personal presentation. Tommaso's attention to personal grooming is documented in abundant detail in *A violent life*. For instance, this concern is manifest in the opening scenes of the novel on a Sunday afternoon when the boys are sporting their best outfits, inspired by new American fashion trends. This is a description of Lello, engaged in a dance at the local communist party recreational center (ARCI).

> Lello chomped chewing-gum and, with a twist of his ass, threw back his thighs tightly sheathed in American jeans, and his feet with their pointed, buckled shoes, first one, then the other.

(Pasolini, 2007:36)

A similar concern is witnessed in a later scene when Tommaso is heading out to join his buddies at the bar.

> To tell the truth, he was impeccable: the sun glistened on the black suit, gilding the rather heavy cloth, as he walked at a calm, controlled pace or as he moved his hand serenely, lifting the cigarette to his mouth. At the end of his pants, his little handsomely-pointed shoes emerged, which he had bought a couple of months ago, but which were still smart.

(Pasolini, 2007:288)

In the long dolly shot of borgata Gordiani in the film *Accattone* the viewer is presented with a series of miserable shacks, but with clean laundry hanging on the line, and poor citizens in pressed shirts and respectable clothing. This is particularly evident in the attire of Accattone's father-in-law, who lives in a condition of diminished humanity yet is dressed in a respectable fashion.

The ideology of civility imposes standards of decency on behavior as well as appearance. Consider the bar scene at Garbatella in *A violent life*, where the owner attempts to subdue a brewing altercation by appealing to an implied concept of civility: 'Come on, [...] we're all Italians! Shake hands, and forget it' (Pasolini, 2007:154). Similarly, in a scene of *Street kids,* in contrast to a general environment of violence and unfair advantage, Riccetto first attacks a young boy in a wanton way, and then, discovering his vulnerability to abuse by a group of older boys, offers his support. While this civic behavior receives no comment, antisocial behavior is instead described through a rich vocabulary. For instance, Pasolini has his protagonists use the expression to act like an American ('fare l'americano') to describe those who do not stand up for their buddies. No adequate English language translation exists, and William Weaver in *A violent life* conveys the idea with the phrase to play dumb: 'Dresser, Smoky and the snotnose, were enjoying the scene, playing dumb ....' (Pasolini, 2007:152). But this is more than playing dumb: it is the denial of innate human character.

If fraternity is expressed throughout these books – with many points of friction – sorority also finds expression. In one scene of *Street kids* a gang of boys are showing off in the failed effort to impress a gang of girls, who show decided capacity to express a common front. In a scene in *A violent life* Tommaso is walking through Garbatella when he encounters a group of young women sitting together in an open field. Women in Italy in those days were subject to rigorous exclusion from public places, at best expressing a muted public presence.[8] Being outside his own neighborhood, and outnumbered, Tommaso chooses not to challenge their momentary control of public space. In an oblique comment, Tommaso vents under his breath resentment that they should so successfully exercise their collective force.

> [Tommaso was] passing alone by the group of girls, staring at them. They pretended not to see him at all, but they caught on right away that he had his eye on them. Then they began to joke more enthusiastically or laugh exaggeratedly, without looking him in the face: and they let him look down below [i.e., under their skirts], as they moved, since they weren't aware of his presence. And besides he was alone, and they were many. Tommasino walked along, one foot after the other, gulping.
> (Pasolini, 2007:81)

Powerful and independent women are seen in many of Pasolini's works, most notoriously in *Mamma Roma*, where the protagonist is ultimately overwhelmed not by gendered violence, but by the sheer brutality implied by her underclass position. Women have the capacity to make a show of collective force even when facing the repressive actions of law enforcement officers, as we see in an episode of *A violent life* when one of Tommaso's friends is threatened with arrest.

The crowd pressed tighter and tighter, especially the women: those already out, and those who lived in the houses near by, who now came out to see. All poor slum women, dishevelled, with black housedresses on, greasy and dirty, slippers on their feet.

The policemen began to shout: 'Clear out! Make room!' But the women who had crowded around wouldn't move; instead, they also began, in voices still a bit subdued, to yell a few words at the flatfeet: 'Shame on you! Bums!'

(Pasolini, 2007:109)

For those who live in the periphery assertive individuality is an important attribute, which must be balanced against the ability to create alliances and friendships, typically among men. Concern for reputation is also prominent. In the final stages of *A violent life* Tommaso says to his buddies at the bar that they should join the effort to rescue the victims of a flood created by a raging storm. They give him a hard time, challenging him to be as good as his word, and go out to provide the aid he proposes. He is trapped by his public assertion, and though completely uncommitted to the rescue effort, he must defend his reputation, and in fact dies in the process. As Pasolini notes, his reputation was his weak point (Pasolini, 2007:308).

Yet notwithstanding strong expressions of solidarity, this is potentially an atomistic society where the threat of violence and unfair advantage is always lurking. In the early scenes of *A violent life* Tommaso bullies younger kids, and continues in a truculent manner until an older brother shows up on the scene, and he idles off with nonchalance (Pasolini, 2007:25). In another scene Tommaso is furious because he has been excluded from a table game at the PCI recreational center, and insults his companions by calling them supporters of the Lazio team, and thus quintessential hicks. Friendship is a barrier against atomism and betrayal. The corollary of this attitude is the rejection of collaboration with authorities, and Tommaso wins himself the nickname of Spy when having been cheated by a friend he heads to the police station to report the fact.

Being shrewd and cunning is a fundamental capacity. In a potentially hostile world cheating others is a form of preemptive self-defense. Tommaso is asked by a soldier on leave to take a photograph of the man and his girlfriend. Tommaso accepts to help, then runs off with the camera, saying the man was such a fool he deserved to be cheated: 'as long as there are dopes a smart kid can always get along' (Pasolini, 2007:104). The original text uses the Roman term micco to signify fool or sucker.

The ability to manipulate others is a positive value, as witnessed when a boy throws a rock for a stray dog to fetch. He throws the rock ever further to wear the animal down, commenting that the dog was a sucker. In Roman dialect then as today to dupe someone is termed turning a person into a subject ('fare soggetto'). The English word tease Weaver uses to convey the idea of 'fare soggetto,' lacks the aggressive implications of the Roman phrase.

Zimmìo [...] was throwing some stones to a dog that had shown up, making him chase them. The animal was exhausted, his coat standing up, his tongue dragging in the dust; he didn't realize they were teasing him ....

(Pasolini, 2007:177)

Many of the forms of behavior described in the Roman periphery are also found in other marginal communities with weak political and economic integration in a wider system. This is seen in a comparison with American inner-city life portrayed in Elijah Anderson's study of Philadelphia's Germantown Avenue. Here too street life is a dominant feature, as is strong patriarchy. Not unlike the Roman periphery of the 1950s, conditions of material and immaterial exclusion shape interpersonal behavior.

In the inner-city environment respect on the street may be viewed as a form of social capital that is very valuable, especially when various other forms of capital have been denied or are unavailable. Not only is it protective; it often forms the core of the person's self-esteem, particularly when alternative avenues of self-expression are closed or sensed to be. As the problems of the inner city have become ever more acute, as the public authorities have seemingly abdicated their responsibilities, many of those residing in such communities feel that they are on their own, that especially in matters of personal defense, they must assume the primary responsibility.

(Anderson, 2000:66)

Assertive individuality is about winning and maintaining the respect of the community, and in Rome this could be achieved by trickery and violence if necessary. In Philadelphia, like in Rome, the personal front is of fundamental importance.

In public the person whose very appearance – including his or her clothing, demeanor, and way of moving, as well as "the crowd" he or she runs with, or family reputation – deters transgressions [...] and may be considered by others to possess a measure of respect.

(Anderson, 2000:66–67)

And, like in Rome, women are subject to strong control by a system which privileges male public presence. Anderson notes the common family breakdown in the American ghetto, something fully evident also in Pasolini's accounts of the Roman periphery. Anderson shows that when the family is intact, a powerful element of patriarchal domination is present.

In public such an intact family makes a striking picture as the man may take pains to show he is in complete control - with the woman and the children following his lead. On the inner-city streets this appearance helps him play his role as protector, and he may exhibit exaggerated concern for his family, particularly when other males are near. His actions and words [signal] that he is capable of protecting them and that his family is not to be messed with.

(Anderson, 2000:39)

The situation in Rome in the 1950s was similar but more extreme, with women exposed to violent forces of patriarchal control (Brownmiller, 1975). Tommaso invites his future girlfriend to the movies, and gropes her during the show so relentlessly that she breaks out in tears. But after the movie, she is portrayed in an image of smug satisfaction; now she has a man she can count on for protection.

Tommaso walked quietly, his hands in his pockets, and Irene came along behind, holding his arm. They walked in silence, like a couple long engaged, who have nothing to do with the rest of the world, all closed in their thoughts ....

(Pasolini, 2007:97)

In Rome of those years the sexual control of women focused strongly on virginity. When Tommaso visits a priest to seek advice for his wedding, the priest, made suspicious by the urgency of the request, asks if 'something had happened,' a colloquialism which can only refer to her virginity. Tommaso is shocked by the suggestion, and responds that she is a nice girl ('una ragazza brava'), respectful of societal norms.

You haven't got into any trouble, have you, with your fiancée? [the priest] asked, 'Has anything happened?' 'Nooo!' Tommaso cried, shocked. 'Nothing like that! You're kidding, father? She's a nice girl!'

(Pasolini, 2007:187)

Many features of personal conduct documented in the Roman periphery have also been documented in Sicily. Jane and Peter Schneider (1976) show how in that southern Italian region concern for respect, strong patriarchy and assertive individuality with its allied antisocial behavior, are all part of a code of conduct adopted to cope with historical marginality. Sicilians are not intrinsically uncivil, rather they are forced to engage in uncivil conduct to deal with violent political and economic domination. This tension between a surface expression of incivility and deep desire for a civil life, is fully evident in Pasolini's descriptions of Rome. This is the force behind the final scene of *Street kids*, when Riccetto stoically represses an emotional response to maintain the appearance of a tough guy. Throughout *A violent life* the reader is confronted with the tension between the desire for a decent life, and the realities of marginalization that make assertive self-help a necessary defensive strategy.

Prolific narrative is devoted to Tommaso's desire to achieve a 'normal' life, with a job, a wife and a family. Thus, he strolls elegantly down Via Nazionale with his girlfriend by his side, looking at the shop windows as might any respectable citizen. But there is no room for him in respectable society. The combination of emotional and material insecurity in an urban environment is nowhere better expressed than in the scene where Tommaso is engaged in altruistic conversation with his neighbor, settling himself with great care against the doorjamb of his shack in an effort to adopt a pose of civil amiability without causing his precarious dwelling to collapse. This is the tenor of the conversation: 'both felt they were sage people, old-fashioned, minding their own business and not looking for trouble' (Pasolini, 2007:126). After their civil conversation, the scene ends with Tommaso struggling to fight back tears as he realizes that the chances of achieving true civility are slim at best.

> And around that wretched pile of huts there was a silence, a peace, a solitude that were frightening. After a little while, without even realizing it, while he stood there alone and downcast, Tommaso felt something like a tear rising in his throat. But he promptly swallowed it again.
>
> (Pasolini, 2007:128)

All of these factors are summed up in an almost didactic way by Pasolini in a scene where Tommaso comes into contact with middle-class kids who are residents of a peripheral neighborhood. This is the INA housing project of Tiburtino IV, that celebrated architectural space built for a community of 4,000 inhabitants (Fig. 3.2). It was built on high ground, with agricultural land all around. The visually dominant feature of the project's dazzling variety of building types are towers seven-stories high, once commanding a vista over most of eastern Rome. Tommaso climbs to one of the upper stories of the most southerly tower and peers out the window, marveling at the vast cityscape which stretches out before him.

> On the landing there was a little oval window, just about the height of his nose. Tommaso went over for a look. From there you could see half of Rome: a pile of houses in the light, the ground already a bit dark, endless; the city seemed to float on the clouds, bobbing up and down, from Montesacro to Piazza Bologna, to San Lorenzo, to Casal Bertone, to Prenestino, Centocelle, Villa Gordiani, Quadraro ...
>
> (Pasolini, 2007:176)

Tiburtina IV was not home only to the street kids we find in all of Pasolini's novels and films on the Roman periphery; there were also middle-class youth. While middle-class families are systematically excluded from Pasolini's narrative, here they receive attention because their presence allows him to characterize the contrast between street life and normal life. The encounter between Tommaso and the respectable

Figure 3.2 Tiburtina IV with its strong sense of enclosure. (Photograph taken in August 2020 © Gregory Smith.)

kids occurs in the recreational center linked to the church. This would have been a DC-aligned ACLI recreational center, which competed all over the periphery with the ARCI centers. The presence of the DC already suggests a claim to bourgeois respectability lacking in descriptions of other neighborhoods. The church which hosted the ACLI was then a simple wooden shack, today replaced by an impressive pagoda style church inaugurated in 1971.

> [The church] was a kind of long, narrow storehouse of pale blond wood, the planks marked with long indentations. The roof was pointed, and at the top there was a cross.
>
> (Pasolini, 2007:180)

The meeting is not entirely successful, partly because Tommaso is a hustler and ill at ease in a setting where he is expected to behave in a civil manner.

> Tommaso was acting like a good boy, a bit jolly and easy, a man after all and, as a man, with his bad habits: gambling, smoking, women…
>
> (Pasolini, 2007:184)

After leaving this meeting, he stops at the fence to observe the middle-class adolescents playing at the ACLI. These kids are clearly different from Tommaso, with an educated style he aspires to emulate.

The kids playing in the church yard must have been students, daddy's boys: they were all Tommaso's new neighbors, more or less.

They were playing, concentrating on the table-ball and ping-pong. They too were dressed smartly, in American jeans plastered with shiny studs, with broad belts and jerseys: but they were all spick and span, soiled a little only on the ass or on their knees, not from work, but because they sat down wherever they felt like it [...]

(Pasolini, 2007:181)

While the middle-class kids are well dressed, Tommaso dons a used suit bought at Campo dei Fiori. His socks have holes he tries to hide by shoving them down into his shoes: 'the crummy socks, stuck a bit too far into the heels of the shoes so you couldn't see the holes' (Pasolini, 2007:185). Using a pretext, he manages to engage the attention of these youngsters, affecting the affable indifference of the well to do. They respond in equal manner in a short vignette where the reader is invited to observe the differences between two conditions of life. The middle-class youth address one another by surname, an official designation of public identity. In Tommaso's circle people address one another using nicknames, like Spy, Shitter, Buddha.[9] Nicknames push away a formal system of institutional control, and suggest involvement in a restricted informal community. The middle-class kids have the support of integrated families, and as daddy's boys have prospects for economic advancement. By comparison to Tommaso, they have a docile temperament, comradely and mutually supportive. Tommaso has a different sort of vitality, and a desperate need to prove himself. His habitual impulse to compete motivates antisocial behavior that can be an asset in this setting. Here is Tommaso's reflection.

If I sign up [with the ACLI], he was thinking, I'll show you how to play table-ball, ping-pong, and the rest! I'll smash you all! In the end, I'll be the big shot around there; what're you and those others anyhow? A bunch of little turds!

(Pasolini, 2007:188)

Ferrarotti recognized belatedly that a condition of marginality could present advantages, and in this context coins the term 'happy poverty' (Ferrarotti and Macioti, 2009). This infelicitous locution defines poverty which is not imposed but accepted. In the Roman periphery of the 1950s poverty was imposed and far from happy. Yet this marginality protected an autonomous form of cultural and human life which Pasolini strove to document and aspired to defend. It was life directly lived, simple and immediate.

In a consumer economy people establish identity through the accumulation of commercial sign values whose meaning is generated by the market system. People with few resources cannot rely on this process, and rather than take meaning from artificial signs, attribute meaning to things with no market value. This is the divide between underclass life and bourgeois experience. As portrayed by Pasolini, a shared understanding of value emerges among the underclass from interaction itself, rather than from a set of sign values predetermined by the market. Community among the underclass entails shifting understandings, subject to the perpetual threat of disruption by the potentially antisocial force of personal vitality. The middle-class community is instead stabilized by an overarching system of social, political and economic control. The collectivity cultivated by the formal system is static and apparently solid. The congregation engendered by spontaneous interaction is fragile and shifting, yet direct and real. The underclass aspires to become a community that it can never fully be, since infused with a powerful element of uninhibited personal assertion that is the foundation for a volatile mutual understanding. Rather than *constituting* a community, Tommaso's entourage is in a perpetual process of *becoming* a community.

Yet though fleeting and unstable, this condition of life provides the immediacy that Pasolini celebrates as vitality. Tommaso notes this when he decides to remain for treatment at the TB clinic, now that he is projected into a world of responsibility and integration. Tommaso is embarked on a new life, and realizes that respectability entails a loss of innocence. From within the confinement of the institutional setting, he peers outside and witnesses evidence of the happiness he had known as a street kid far removed from any form of integration. In this scene his new found friends, activists in the PCI, have climbed over the fence and are escaping from the clinic.

> Tommaso watched them run across the street: they reached the other side, near the mechanic's, and headed for the bus stop: all around them there was a bustle of traffic and people at the supper hour. From some old tenements a troop of little kids came down toward the stop, heading for God knows where.
>
> Their faces dirty under their forelocks, they walked arm in arm, all talking eagerly, paying no attention to anyone. Some talked on and on, others just laughed. And those little smartass faces, above the filthy, colored collars, were the very image of happiness: they didn't look at anything, and they went straight where they had to go, like a herd of goats, sly and carefree.
>
> Aaaah, Tommaso sighed, I was rich and didn't even know it!
>
> (Pasolini, 2007:250)

In these scenes Tommaso is attracted to the PCI owing to the promise of an enduring altruistic society. Yet later, when he goes to take out a party card, he discovers that the PCI activists are as self-serving as anyone, and simply use the party as a vehicle for personal advancement. Tommaso's signs up anyway, in the vain hope that some collective advantage might be found through party membership.

> And so it was: a few days later, Tommaso turned up at the section, with the two people who had to be his witnesses, Delli Fiorelli himself and Hawk; he was signed up, paid what he had to pay, and finally he was able to have his slice of the pie: he put the card in his pocket, ready to struggle for the red flag, too.

> (Pasolini, 2007:271)

The failure of the party to create an altruistic community in the periphery may be attributable to the nature of community itself. The party's ideal was a bourgeois community stabilized through control by an external force. A real community – that of the underclass – comprises citizens beyond control, sly and carefree. The process of hypostatization advocated by the party erases the underclass's spontaneous fleeting character. The plenitude of happiness Tommaso describes as a street kid, is life lived in a condition where meaning is generated through spontaneous interaction. The paradox is that to develop this society in economic terms, using the model advocated by the PCI, requires the formation of a new type of community and the attendant betrayal of the underclass's intrinsic vitality. This is the implication of Pasolini's reasoning.

Seen from this perspective community in the periphery is an aspiration rather than a reality. In the center it is instead a hypostatized fact. We are dealing with two distinct urban conditions, different from a sociological standpoint and as physical locations. In the center we find well-curated spaces reflecting powerful mechanisms of control. Peripheral spaces instead lack abiding overarching logic. Life in the center is united by a comprehensive design, while in the periphery each element is disconnected from and in some way reciprocally opposed to all others. In the center we find the hierarchical ordering principle typical of bourgeois society. In the periphery we have an assemblage of multiple parataxic forms. In this way the aesthetics of the community and the aesthetics of the city mirror one another.

Can we theorize a culture of marginality? Pasolini believed so. In particular he held that the marginality of the urban poor was also that of the rural south. He was especially fascinated by Naples.[10] Though not rural it was southern, a preeminent expression of that pan-southernism which he believed united all people from the global south in a condition of pure humanity.

Comparisons with ethnographic explorations of Naples in the 1950s

Comparison with ethnographically inspired psychoanalytical studies of poor Neapolitan communities in the late 1950s provides analytical insights into the life of the Roman underclass in this period. In Naples we see that conditions of marginality shape not only external codes of conduct, but also impact deeply situated personal identity. Anne Parsons was a social critic and political activist, as well as a participant observer (Breines, 1986). She worked in the late 1950s with patients and their families in the psychiatric hospital of Naples, visited homes, interviewed several families and produced two studies that are relevant to our pursuits. One was published in 1964, the year before her premature death, entitled 'Is the Oedipus complex universal?' The other was published in Italian in 1962, under the title 'Paternal and maternal authority in the Neapolitan family.'[11]

Her findings mirror Pasolini's portrayals, with strong gendering of public space, and a dramatic cleavage in the image of women, torn between the negatively idealized prostitute, and the positively idealized mother. The positively idealized mother is selfless, and her love unconditional. All the mothers in Pasolini's accounts fit this description. In one scene Riccetto, blinded by a fit of unreasonable fury, attacks his mother with a knife. She is wounded, but plays down the episode to avoid criminal consequences for her son. In another scene, when Tommaso returns home from prison, it is his mother who greets him, in tears, and provides the material support needed to re-establish himself. Fathers are part of a public world, removed from the home. They have responsibility for ensuring the material support of the family, but little more. The separation between the world of men and women is complete, Parsons tells us. Men conduct their business as they see fit, and women are expected to refrain from raising questions about activities outside the home.

Parsons describes powerful anticlericalism in her Neapolitan community. In *A violent life* the church is also negatively idealized. The priest interacting with Tommaso when he seeks information about his marriage is portrayed most unsympathetically. He is distant from the concerns of his flock, obsessed solely with the defense of the moral order which validates his authority.

> The priest [...] must have been sick, because his skin beneath the sparse beard was white and grey, his eyes hollow, his lips pale as a cat's. He was small, and so thin he seemed lost inside his cossack.
>
> (Pasolini, 2007:185)

In another scene, the parish priest lies publicly about a chicken theft involving four hens.

> [The priest said] that they stole thirty hens from him last night. That blasphemous thieves had broken into the henhouse, and that these sinners had taken thirty hens, exploiting him, who lives only for charity. Thirty hens, he said, the sonofabitch!
>
> (Pasolini, 2007:105)

In Naples, swearing is characteristic of boys, as is anticlerical joking. A common curse in Roman dialect is to swear against saints, and even against God himself.[12] As in Rome, adult male life takes place on the street corner, or at the bar. Social interaction is humorous and partly cynical. The attitude of boys to authority, both secular and religious, is more negative than positive. Street life is shunned by woman: the street is inherently sexualized and dangerous. Women in public are considered prostitutes and fair game for abuse. Young Neapolitan girls become very excitable and giggly when they have the occasion to be in public, and street phobias are common. Decent women are associated with the home.

Parsons notes that family life is beset by aggression and conflict. This is partly caused by strong gender segregation and the absence of stable work opportunities. Women in fact often work, but this activity undermines the stability of the male role and thus receives no public support. Family should compensate for the lack of institutional and economic support, yet fails to do so. All of these factors combined lead to violence which is also seen in Rome: 'After a while the door opened again and Alduccio's father came in. He was drunk, as he was every night. He approached his wife and started to beat her' (Pasolini, 2016:176). In Naples as in Rome women have ongoing relations with the natal family and neighbors. When Tommaso first meets his future girlfriend, she is wearing her housedress, assembled in a crowd of neighbors in front of her housing project. Likewise, in the scene where Tommaso walks past the group of young girls sitting in an open field, the girls gain strength by being neighbors.

For men instead needs for comradeship and mutuality are provided by the extrafamilial male peer group. The division of sexes is such that marriage is not the focus of reciprocal affective exchange. The major requirement for the husband is to feed and support his family, and for men a job offers rewarding proof of masculinity. The paradox in Naples as in Rome is that men must provide food and drink for their male peers to gain status, and this removes food from women and children. Parsons notes that within the Neapolitan home there is no space for a ritualized meal, no regular eating patterns. This would appear to be a typical meal in the Roman borgata.

> When lunch was ready, Tommasino ate a few spoonfuls of soup in a hurry, took his bread with some greens in it, and went outside again, chewing.
>
> (Pasolini, 2007:27)

In Naples male rage is common, and men are often unwelcome in the family. The handling of children is rough, with a distinctive element of aggression. The father is severe and distant, treated with respect rather than intimacy and affection. Yet there is a high degree of dependency on parents. This analysis of the Neapolitan family is mirrored in Pasolini's literary accounts.

The best illustration of this family situation is seen in Tommaso's dream, which reveals a frustrated desire.

> Suddenly, from one of the little paths among the shacks Tommaso's father appeared [...] 'Tommaso, you had your breakfast?' he asked Tommaso, as he came up. Tommaso looked at him in amazement, because it was the first time in his life his father had asked him that question.
>
> (Pasolini, 2007:220)

As the outcome of her careful analysis of Neapolitan families, Parsons concludes that the type of family structure valued in southern Italy is not viable in conditions of poverty. The patriarchal model cannot be maintained without economic opportunities which allow the father to be considered the family provider (Parsons, 1969:xiv). This remarkable finding coincides with Pasolini's description of family pathology in peripheral Rome in the same period, a description which would have shocked many readers at that time. The family described by Parsons has been defended by the Catholic Church for centuries, and patriarchy was not only defended by the church and its political partners in the DC, it was also supported by the PCI. The party was anxious not to antagonize the church, and thus accepted to leave family matters in the hands of the DC (Bernini, 2008). Late and weak support for gender equality in Italy is partly due to the PCI's exclusive focus on class struggle. Italian feminists eventually rebelled against this idea, but only in the 1980s when they called for a multipolar dialectic which involved gender as well as class struggle (Birnbaum, 1993:258). Pasolini's quest in the 1950s for a Marxism of the sentiments somehow echoes this call, and shows just how distant he was from the party establishment. Pasolini's critique was not only against what he called the 'sexual racism' of Italians in regard to homosexuality, it was also an implicit critique of the broader gender violence upon which the edifice of patriarchal morality rests (Siciliano, 1978:231).

The members of the underclass faced massive odds in achieving a full personal life. Yet stark adversity notwithstanding, these marginal citizens expressed a sense of vitality that fascinated Pasolini. This was the unexpected piety described by Carlo Bo.

(Mild) poetic liberties

Both novels furnish Pasolini with the opportunity to exhibit his knowledge of the exact spatial relations which characterize the various locations where narrative action takes place. We see in his accounts that residents of the periphery frequent all parts of the city, engaging with a wide range of people and places. The sequence that especially illustrates this spatial mastery is the episode in *A violent life* where Tommaso starts from his shanty at Pietralata, and then journeys through the center in various encounters. He travels out to the countryside, and back into the city. He then heads back to Pietralata.

The action starts in Via Nomentana, presumably at the intersection with Via Pietralata which leads up to Via dei Messi d'Oro, and the chaotic open landscape which once housed the shanty then called Shanghai. The shanty was about 500 meters from Pasolini's home at Rebibbia in 1951. Tommaso and his buddies board the bus that leads down Via Tiburtina, and eventually reaches Via di Ponte Quattro Capi, right off the Jewish Ghetto where still today various city buses have their terminus. The ghetto was of course the location of Pasolini's first residence in Rome. From there they traverse the two bridges that connect the Tiber's left bank to the right, crossing over Tiber Island. On the right bank of the Tiber they reach the beginning of Via della Lungaretta, a fairly straight road which leads west on the way to Saint Peter's. They are now in Trastevere, a district traversed more or less north to south by a road dating to the 1883 master plan, called King's Road (Viale del Re) until it was changed to Viale Trastevere in 1949. Pasolini still calls this King's Road. Here Pasolini notes activity around the Reale and Esperia movie theaters. Right behind the Reale they reach Vicolo della Luce where there was once a party cell belonging to the right-wing group, MSI. Here they meet a fascist friend, and head back to Tiber Island, stopping to buy olives from a street vendor at Piazza dei Ponziani. Other friends join them here, coming down Via dei Vascellari which intersects the piazza, before they reach the Tiber again at Ponte Rotto.

Here they cross the Tiber, and follow an unstated itinerary which eventually leads them to Largo Argentina. There they join a gang of fascist sympathizers coming from different parts of the city: Borgo Pio, Ponte, Monteverde and Alberone. From Largo Argentina they move east toward Palazzo Altieri, then a few blocks north to Piazza Minerva. They then pass behind the Pantheon, and up to Piazza della Rotonda. The police come from the east along Via del Seminario, and from the south starting at Piazza Minerva, to break up the fascist gathering, and the friends escape by heading to the west. They follow Via del Teatro Valle, until connecting with Via del Governo Vecchio which leads back to a higher reach of the Tiber. At the Vittorio bridge they board a tram and head back to where they started, getting off at the Garibaldi bridge and crossing over to Viale Trastevere. The tram is the 'circolare' which ran along the Aurelian Walls, then along the Tiber embankment road before returning to the walls at a southern point. The friends have a pizza at a restaurant still there today, and then continue their adventure. This includes the theft of a car at Largo Argentina, and a journey that leads them through the countryside surrounding Rome.

All of these relationships are exact, and still today can be followed by anyone interested in doing so. Pasolini is showing off his perfect knowledge of the city, building its fabric into the structure of his narrative. But at a certain point he takes minor liberties to build a sense of excitement. This is when the street kids have stolen a suitcase from a German pilgrim out on the eastern side of the city, at Torpignattara, and have come back to Trastevere where they sell the stolen goods to a fence. The mild poetic license begins when he has the protagonists drive across Ponte Sisto from the Trastevere side and then turn right into Lungotevere.[13]

All of these details are accurate. After traversing Ponte Sisto at high speed, they reached Ponte Rotto, then Ponte Sublicio, and then the Ostiense train station, each in three seconds one from the other. Yet the last leg of the journey is one and a half kilometers long, impossible to achieve in three seconds at any speed by car.

> 'Let's goooo!' Tommaso yelled. Lello turned here and there [travelling through Trastevere], at random, into some alleys, then over Ponte Sisto and along the Lungotevere, speeding. The rain had stopped, there were patches of clear sky, brightening now. In three seconds they were at Ponte Rotto, in another three at Ponte Sublicio, and in three more at the Ostia station ....
>
> (Pasolini, 2007:55)

This desire to create dramatic tension certainly does not diminish the ethnographic value of the text. As novels, these books take some liberties with factual representation, drawing the reader to deeper psychological involvement. In *The method* he notes that his research has two aims, one naturalistic-documentary and the other sensual-stylistic. The first provides raw material woven into the second, creating narrative tension designed to engage the reader.

Similar liberties were later explored in his films on Rome, both in *Accattone* and *Mamma Roma*. Rhodes sees these liberties as a deliberate effort to establish a distance between the viewer and the scenes portrayed (Rhodes, 2007:55). This is especially seen in the dolly shot in borgata Gordiani, portraying spatial and temporal relations which are impossible in the real world. This editing technique disorients the viewer, and reminds them that they are viewing a representation of the world. We also see this in the final episode of the film *Accattone*, where the underclass protagonist steals a motorcycle and attempts to escape arrest. The background scene allows the viewer to see the flight start in Testaccio, but then ends with Testaccio visible on the other side of the Tiber, with no scene showing the passage from one bank to the other. These modernist techniques, Rhodes argues, are used against the modernism Pasolini associated with the reformist policies he rejected (Rhodes, 2007:55). The films are realistic, but move beyond pure realism.

The periphery today

The underclass of Pasolini's day no longer exists. To be sure, Rome still has a significant poor population. The Vatican relief agency, Caritas, continues to operate and provides 11,000 free meals to needy citizens every day (Manna and Esposito, 2019). Another Catholic relief agency which also provides free meals, Sant'Egidio, reports that a majority of their patrons are foreign.[14] We still have homeless citizens, about 20,000 according to Caritas. Poverty is prevalent especially among elders and families with young children. They are not homeless, but they are poor. Caritas organizes over 50,000 charity visits each year to the homes of needy elders. They estimate that about 6% of Rome's elders are poor, with peaks of 12% especially on the eastern fringe (Manna and Esposito, 2019).

We still witness family pathology, and violence against women is a central topic in Italian political debate. In 2018 a woman was murdered in Italy every 65 hours, and almost 80% of femicides were committed by a former or current partner, or some family member.[15] Prostitution has been essentially unregulated since the Merlin law of 1958. According to some estimates, 50,000–70,000 prostitutes are active in Italy today, and upward of nine million Italian men use their services (Scaglione, 2013). The geography of prostitution in Rome is public knowledge, and the service is available at all times of day and night. It is evident that heterosexual intimacy continues to be governed by forces present in the 1950s. Homosexual prostitution is also common, and even though civil unions are allowed among same-sex couples, homosexuality is still taboo in wider society.[16]

The physical characteristics of the periphery have also changed since the times described by Pasolini. Tiburtina IV, for instance, is completely engulfed by mostly middle-class midrise buildings dating to the 1970s. The flats in the housing project itself have been sold privately, and are highly desirable on the real estate market. The Pamphili housing project has similar appeal, and here too units are being sold as private real estate. Today the periphery is chiefly comprised of middle-class homes and families, the bourgeoisie whom Insolera claims always dominated the city anyway.

It is evident that many things have changed, but what persists is a powerful idea of the dignity of life in the periphery, beyond the privileged circle of the center. This may be the most important facet of Pasolini's heritage. Even progressive Catholics decry the materialism of modern-day life, namely, the logic of the consumer market. This is the logic Pasolini associated with the 'development' of society: the production of ever more commodities of no real utility. It is different from the 'progress' of society as a civic concept championed by Pasolini. Italy performs well when it comes to measures of progress along this scale. An anodyne example is the significant presence of volunteer workers throughout the country. An impressive one in ten Italians do volunteer work organized by over 40,000 associations.[17] These citizens are looking for something that goes beyond the embrace of material sign value. They seek deeper expressions of humanity, and membership of a more meaningful congregation. Perhaps they are looking for a more profound sense of piety.

Notes

1   I make this observation based on many years of using selected Pasolini writings in undergraduate courses held at the University of California Rome Center on the sociology of Rome. The novels in particular provide revealing insights into conditions of life in the periphery in the 1950s.
2   Signorelli claims that while de Martino defined himself as an itinerant ethnographer of the Italian south, he was superficial in describing his method of investigation (Signorelli, 1986:5). This makes Pasolini's detailed concern to document his research method all the more surprising.
3   De Martino's *Sud e magia* ends with the hope that the peasants of the south would eventually abandon 'their sterile embrace with the cadavers of their history, to open the way to a higher and more modern heroic destiny' (de Martino, 1959:139). The imperative to modernize backward peasants was a principle Pasolini rejected. He believed that peasant culture possessed desirable qualities unrecognized by his contemporaries.
4   Pasolini uses the term *ragazzi del popolo*. We must note that the word *popolo* in this context embodies a strong class dimension, suggesting membership of a subaltern group opposed to a hegemonic one. It is difficult to translate into English this concept of separation from and resistance to a dominant system of political and economic control.
5   Siciliano reports that they used to dine together regularly in a pizza restaurant near borgata Gordiani. At that time Sergio earned more money than Pasolini and paid for the meals (Siciliano, 1978:188). Later Sergio Citti became a noted filmmaker thanks to Pasolini's support.
6   In an extended period spent surveying the territory where the Pamphili project is located, in spring of 2013, I asked a local citizen who had known Pasolini in the 1950s whether the poet engaged in any of the activities that made him so controversial in Friuli. The citizen was adamant in stressing that Pasolini's behavior in their community was beyond reproach, confirming the statement made in *The method*.
7   https://blog.zingarate.com/condividilacalabria/quanti-calabresi-ci-roma/ (Accessed November 12, 2019).
8   For the muted public role of women see Shirley Ardener (2005).
9   Others include Smoky, Dresser, the American (Pasolini, 2007:150); Hyena, Prickhead (177).
10 Pasolini believed that the people of Naples lived in an historical void putting them in direct contact with their archaic human condition (Plastino, 2008:50).
11 Both essays are included in Parsons (1969).
12 Pasolini renders this as '*Porco d...*', meaning *porco dio*, a common Roman imprecation against God.
13 In the 1950s vehicle traffic moved in both directions along this stretch of the Tiber embankment road. Today traffic only moves upstream on the river's left bank.
14 https://www.santegidio.org/downloads/Con_i_poveri_a_Roma.pdf (Accessed November 12, 2019).
15 https://www.istat.it/it/violenza-sulle-donne/il-fenomeno/omicidi-di-donne (Accessed November 12, 2019).
16 https://www.osservatoriodiritti.it/2019/06/04/omofobia-significato-in-italia-tema-legge/ (Accessed November 12, 2019).
17 https://www.aynicooperazione.org/uno-sguardo-al-volontariato-in-italia/ (Accessed November 12, 2019).

Bibliography

Anderson, Elijah (2000) *Code of the street. Decency, violence, and the moral life of the inner city*. New York: W.W. Norton & Company.

Ardener, Shirley (2005) 'Ardener's' "muted groups": the genesis of an idea and its praxis,' *Women and Language*, 28:2:50–54.

Asor Rosa, Alberto (1979 [1965]) *Scrittori e popolo. Il popolismo nella letteratura italiana contemporanea*. Rome: Savelli.

Berlinguer, Giovanni (1955) 'Il vero e il falso delle borgate di Roma,' *L'Unità*, 29 July.

Bernini, Stefania (2008) 'Family politics: political rhetoric and the transformation of family life in the Italian second Republic,' *Journal of Modern Italian Studies*, 13:3:304–324.

Birnbaum, Lucia C. (1993) *Liberazione della donna: feminism in Italy*. Middletown: Wesleyan University Press.

Bo, Carlo (1955) 'Il romanzo dei ragazzi disperati delle borgate romane,' *Europeo*, 19 June.

Bravo, Gian Mario (ed.) (1965) *Annali franco-tedeschi: Arnold Ruge e Karl Marx*. Milan: Edizioni del Gallo.

Breines, Winifred (1986) 'Alone in the 1950s: Anne Parsons and the feminine mystique,' *Theory and Society*, 15:6:805–843.

Brownmiller, Susan (1975) *Against our will. Men, women and rape*. New York: Simon and Schuster.

Cagnetta, Franco (1954) 'Inchiesta su Orgosolo,' *Nuovi Argomenti*, 10, Sept-Oct.

Cagnetta, Franco (2002) *Banditi a Orgosolo. With biographical notes by Paola d'Errico*. Florence: Guaraldi.

Chiarcossi, Graziella and Franco Zabagli (2017) *La biblioteca di Pier Paolo Pasolini*. Florence: Gabinetto Scientifico Letterario G.P. Vieusseux.

de Martino, Ernesto (1959) *Sud e magia*. Milan: Feltrinelli.

Fellini, Federico (dir.) (1957) *Le notti di Cabiria*. Rome: Dino de Laurentis.

Ferrarotti, Franco (1979 [1970]) *Roma da capitale a periferia*. Bari: Laterza.

Ferrarotti, Franco and Maria Immacolata Macioti (2009) *Periferie. Da problema a risorsa*. Rome: Sandro Teti Editore.

Gadda, Carlo Emilio (1957 [1946]) *Quer pasticciaccio brutto de via Merulana*. Milan: Garzanti.

Lizzani, Carlo (dir.) (1960) *Il Gobbo*. Rome: Dino de Laurentis.

Manna, Elisa and Raffaella Esposito (eds.) (2019) *La povertà a Roma. Un punto di vista. Anno 2018*. Rome: Caritas Roma.

Merlin, Lina and Carla Barberis (1955) *Lettere dalle case chiuse*. Milan: Gallo.

Montanari, Massimo (2009) *Il formaggio con le pere. La storia di un proverbio*. Bari: Laterza.

Parsons, Anne (1969) *Belief, magic and anomie. Essays in psychosocial anthropology*. New York: The Free Press.

Pasolini, Pier Paolo (dir.) (1961) *Accattone*. Rome: Cino Del Duca.

Pasolini, Pier Paolo (1962a) *Il sogno di una cosa*. Milan: Garzanti

Pasolini, Pier Paolo (dir.) (1962b) *Mamma Roma*. Rome: Arco Film.

Pasolini, Pier Paolo (dir.) (1963) *La ricotta*. Rome: Arco Film.

Pasolini, Pier Paolo (dir.) (1966) *Uccellacci e uccellini*. Rome: Arco Film.

Pasolini, Pier Paolo (1979) *Ragazzi di vita*. Bari: Einaudi.

Pasolini, Pier Paolo (1986) *Lettere 1940-1954 con una cronologia della vita e delle opere*. Edited by Nico Naldini. Bari: Einaudi.

Pasolini, Pier Paolo (2006) *Accattone, Mamma Roma, Ostia*. Milan: Garzanti.

Pasolini, Pier Paolo (2007) *A violent life*. Translated by William Weaver. New York: Carcanet.

Pasolini, Pier Paolo (2016) *Street kids*. Translated by Ann Goldstein. New York: Europa Editions.

Plastino, Goffredo (2008) 'Introduzione,' to Alan Lomax, *L'anno più felice della mia vita. Un viaggio in Italia 1954-1955*. Edited by Goffredo Plastino. Milan: Saggiatore.

Putnam, Robert D. (1993) *Making democracy work. Civic traditions in modern Italy*. Princeton: Princeton University Press.

Rhodes, John David (2007) *Stupendous, miserable city: Pasolini's Rome*. Minneapolis: University of Minnesota Press.

Scaglione, Fulvio (2013) 'Liberi i clienti, schiave le prostitute,' *Famiglia Cristiana*, 29 November.

Schiaffini, Alfredo (1954) 'Introduzione,' to Leo Spitzer, *Critica stilistica e storia del linguaggio*. Edited by Alfredo Schiaffini. Bari: Laterza.

Schneider, Jane and Peter Schneider (1976) *Culture and political economy in western Sicily*. Cambridge, MA: Academic Press.

Siciliano, Enzo (1978) *Vita di Pasolini*. Milan: Mondadori.

Signorelli, Amalia (1986) 'Lo storico etnografo. Ernesto de Martino nella ricerca sul campo,' *La Ricerca Folklorica*, 14:Apr:5–14.

Silverman, Sydel (1975) *Three bells of civilization. The life of an Italian hill town*. New York: Columbia University Press.

Smith, Gregory (2013) *La comunità e lo Stato. Antropologia e storia nella Marsica del Novecento*. Luco dei Marsi: Aleph Editrice.

Spitzer, Leo (1954) *Critica stilistica e storia del linguaggio*. Edited by Alfredo Schiaffini. Bari: Laterza.

Tonelli, Anna (2015) *Per indegnità morale*. Bari: Laterza.

Wellek, René (1960) 'Leo Spitzer (1887-1960),' *Comparative Literature*, 12:4:310–334.

4 Pasolini, the Roman periphery and the sacred

The sacred

A milestone in Pasolini's poetical production is the collection of poems published under the title *Gramsci's ashes* in 1957. This was a pivotal point in his reception by the general public, and the first edition of the volume sold out within a month (Siciliano, 1979:212). These poems also document the transition from his cultivation of the myth of Friuli to his fascination with the Roman underclass. The editorial success of the collection may be linked to the troubled times faced then by progressive forces in Italy. 1956 had been a doubly traumatic year for communists, with on the one hand the discovery of the horrors committed under Stalin, and on the other the invasion of Hungary by Soviet forces. Both events shocked many PCI supporters, leading to the loss of some 200,000 party card holders within the period of a year (Agosti, 2012). These circumstances caused supporters to question established certainties, and look for new frontiers, including the writings of a dissident communist like Pasolini.

The collection is also a turning point in Italian literature, credited with having terminated the course of Italian Neorealist poetry (Turconi, 1977). Realism is a fundamental current in European literature with roots in the 19th century. It eventually gave way to Neorealism in widely different circumstances with much debate as to origins and definitions. While only a handful of poets are associated with Neorealism, the movement also includes a large number of novelists, reaching back to Alberto Moravia's *Gli indifferenti* (Moravia, 1929), and a significant number of filmmakers, starting in 1945 with Rossellini's *Roma città aperta* (Rossellini, 1945). Pasolini is generally associated with this literary and filmic movement, poised between Neorealism and later developments, always with a strong realist vein.

Neorealist poetry devotes concerted attention to the contemporary world. Just as Neorealist films are shot on location, so Neorealist poetry has a distinctive spatial orientation. Place guides poetic inspiration. The main spatial axis organizing *Gramsci's ashes* is the shift from Friuli to the Roman periphery, with scattered treatment of other Italian locations, especially the south and the poor mountain communities from which many dwellers of the Roman periphery had migrated. So strong is the spatial dimension in Pasolini's work that it endows place with a sacred character (Verbaro, 2017). Mircea Eliade (2008) was a primary source for this treatment, seeing the world's magical quality as being engendered by profound immanence (Felice and Gri, 2013). Eliade calls this quality hierophany, a sacred dimension integrated with the profane, giving human existence direction and purpose. There is nothing transcendental about hierophany: it is all about being in the physical world.

As in many of Pasolini's works, the collection *Gramsci's ashes* explores the way the forces of modernity undermine an archaic sacred condition. His poem about Friuli, 'Quadri friulani,' sings the praises of the simple peasant life, with its 'allegria presente … sereno futuro' (happy present and serene future). His poem 'L'umile Italia' (Humble Italy) celebrates the past life of simple rural folk living carefree in an enchanted world. Enchantment persists in the Roman periphery, and in 'Il canto popolare' (Folk song) he describes a young man living in Rebibbia 'cantando, l'antica, la festiva leggerezza dei semplici' (singing the ancient, festive light-heartedness of simple people). This sacred simplicity is haunted by the anticipation of an ominous turn of events ready to smother the idyll. 'Comizio' (Political rally) alludes to the desperate quest for a ray of sunlight in a new world born on a dark morning. As a whole, the poems probe the sources of the sacred, and show how the circularity of the human condition allows for the glimmerings of the sacred light to pierce through contemporary obscurity. They describe a natural world which is never just natural: it is the mystical totality of the sacred (Verbaro, 2017:77).

The poems identify a bridge between the sacred opulence of the past and the uncertain course of the future (Verbaro, 2017:93). The bridge is recognition of the primordial qualities preserved in contemporary time and space. In Rome the contrast between the sacred and the profane, the archaic and the modern, corresponds to the distinction between the periphery and the center. These incommensurate realms are inextricably interconnected owing to Rome's syncretic character, plural and heterogenous, definitionally total. Periphery and center form an oxymoronic unity. The roots of the sacred run especially deep in the periphery, masqueraded in representations of the center as a wasteland. Yet notwithstanding the power and prestige of the center, Rome in its totality still expresses a spontaneous sacred quality.

Geographical sensitivity is part of a modern turn in the social sciences, an area of increasing interest aiming to enhance our ability to live completely. Indeed, attention to physical place is the subject of various modern disciplines, the most relevant from our standpoint being ecocriticism and geocriticism. Iovino (2016) has written magnificently about the relationship between literature and geographical experience in Italy. Her ecocriticism describes the agency and textuality of nature, the reciprocal permeability of matter and imagination. The individual is part of a totality which expresses reciprocal resistance and liberation: human subjects and the physical world both have agency. They form together a transcorporeal assemblage balancing the porous quality of human subjects with the permeable world around them. Landscapes and bodies alike are texts, and as with Pasolini, they each tell a story.

Another discipline of obvious relevance to our investigation is geocriticism, especially Edward Said's (1985) study of Orientalism. Although the specific origin of Orientalism is geographically remote from Italy, the term documents a process whose pertinence to Italian history and society has been the subject of much scholarship (e.g., Schneider, 1998). Said indicates that Orientalism is 'a Western style for dominating, restructuring, and having authority over the Orient' (Said, 1985:3). It is a western concept which impacts definitions of both east and west. There is strong similarity with the Italian case, especially in the relationship between north and south, center and periphery. Orientalism in Italy is also used to dominate and control. Yet in Pasolini's terms, it is not merely a portrayal of difference, it is the misconstrual of a true distinction. The south is incommensurately different from the north. A distorted image allows the center to wield control. The way to free the south, and along with it the periphery, is to acknowledge their special character. As is generally the case, Pasolini does not clarify this thought in exposition, instead he suggests this interpretation especially through his poetry. This understanding is evident in two poems contained in the collection *Gramsci's ashes*. One is the title poem, 'Le ceneri di Gramsci' (Gramsci's ashes), the other is what I consider to be a recusant ode, 'Il pianto della scavatrice' (The tears of the excavator).

'Gramsci's ashes' is written in Pasolini's corporeal style. The founder of the Italian Communist Party had died in a fascist prison in 1937, and his living body was unavailable for poetic treatment in the 1950s. But his tomb was within reach, ensconced in Rome's a-Catholic cemetery abutting on an ancient funerary monument taking its appearance and name from a pyramid (Fig. 4.1). This location marks the passage from the historic city to the periphery, following the trajectory of Rome's proletarian growth in the early 1900s (Fig. 4.2). It is a rather exclusive place, dedicated to British and other non-Catholic expatriate communities. Among its illustrious guests are John Keats and Percy Bysshe Shelley. Gramsci was accorded the honor of burial owing to his marriage with Julia Schucht, whose Orthodox Russian family had built their tomb there. No Schucht ever used the tomb, so Gramsci is alone in using the sepulcher, treated as a non-Catholic expat in Rome (Beck-Friis, 1989).

Pasolini is a master in using the cyclical character of the sacred to fuse together past and present, fashioning Italy's extraordinary literary and artistic achievements to novel temporal requirements. He gives substance to Eliade's claim that hierophanies represent an eternal process of becoming (cf. Mariotti, 2007:147). One of the most celebrated poems outlining this technique is composed in diary form as part of a collection called *Poesie mondane* (*Everyday poems*). These were conceived as a diary reflecting everyday

Figure 4.1 Gramsci's tomb in Rome's a-Catholic cemetery. (Photograph taken in April 2020 © Gregory Smith.)

thoughts. The poem dated June 10, 1962 is about Rome and Pasolini's love for the past. This short excerpt captures the poem's temporal musing.

> Io sono una forza del Passato.
> Solo nella tradizione è il mio amore.
> [I am a force of the Past.
> Only in tradition is my love.]
>
> (Pasolini, 2012:24)

But Pasolini is no antiquarian, and he goes on to state that he is 'more modern than anything modern,' ending the poem by lamenting that he is an

Figure 4.2 The Pyramid at Porta San Paolo seen from the south. The a-Cemetery is located to the left of the Pyramid. (Photograph taken in August 2020 © Gregory Smith.)

adult fetus cut off from its roots, wandering through a forgotten landscape. Such is the fate of culture in the wasteland of modern times. This poem achieved prominent exposure owing to the fact of having been read aloud in a critical scene of the film *La ricotta*.

This celebration of the past in a present context is an operation he carries out in a similarly explicit way in the poem 'Gramsci's ashes.'[1] So transparent is this poem's tribute to one of Italy's greatest poets – Giacomo Leopardi – that Asor Rosa with his usual caustic turn claims that it is a 'Marxistized' rewrite of Leopardi's celebrated poem 'Silvia' (Asor Rosa, 1979:492). The comment is calibrated to discredit Pasolini, but in fact reflects precisely the modern poet's desired objective. Leopardi was far from being a political activist, but the immediacy of his poetic style and his deep civic sensitivity attracted Pasolini, as did the stylistic devices which anchor experience to physical location. This is specifically in relation to the use of deictic allusions binding poetry to place (Banda, 1990:182). While diexis is usually restricted to literary analysis, it has eminent geosemiotic relevance, linking as it does language to the world outside of language (Scollon and Scollon, 2003:209). A brief passage of Pasolini's poem is sufficient to illustrate the point.

> […] in questa magra serra,
> innanzi alla tua tomba al tuo spirito restato
> quaggiù tra questi liberi [...]
> [... in *this* sad enclosure,
> before *your* tomb in the *presence* of *your* spirit
> *here* among *those* who still live free...]

(Pasolini, 1976:60; italics added)

Reading this poem places us in a cemetery, physically surrounded by an abundance of tombs. Shelley's internment a few meters from Gramsci is a physical coincidence allowing Pasolini to reveal the deeper aim of his opus. Historically this was a critical moment for the communist movement in Italy, surrounded by doubts concerning Soviet communism's historical pathway. It is here that Pasolini takes his distance from the tenets of Marxism, in particular the rational dialectics of class struggle which should propel society into a bright future through the force of history and reason. The poem speaks of a contradiction, where he is with Gramsci in his heart, but against him in his aesthetic passion for a proletarian life which far predates the writings of his Marxist master (Pasolini, 1976:71). Passion, not reason, guides Pasolini's thought, as well as the foreboding that humankind's archaic vitality is destined to disappear. This thought justifies the poetic pretext of citing Shelley as a 'vortex of sentiments' moved by 'the carnal joy of adventure, aesthetic and puerile,' a vitality which prefigures Shelley's death by drowning. The poem goes on to ask Gramsci whether he, Pasolini, can live without the desperate vitality which had attracted the ferocious censure of the PCI as being a degenerate aesthete.

> Mi chiederai tu, morto disadorno,
> d'abbandonare questa disperata
> passione di essere nel mondo?
> [Will you ask me, unadorned dead man,
> To abandon this desperate
> Passion for being in the world?]
>
> (Pasolini, 1976:75)

Shelley was well suited to inspire Pasolini's particular radicalism. Incensed by the violence used against ordinary working men and women, Shelley exercised his poetical powers to denounce injustice and galvanize action.[2] Karl Marx had read Shelley's poetry, and allegedly claimed that 'Shelley was a thorough revolutionary' (Foot, 1981:227). Yet Shelley's poetry also had the power to ponder the sublime character of the world, allowing him – like Pasolini – to move between political invective and the poetic celebration of physical experience (Leighton, 1984).

The other poem we must consider here is what I call a recusant ode using a poetic form made popular in the works of the British romantic poets. This is Pasolini's 'The tears of the excavator,' celebrating not a classical antique artifact but a prosaic feature of the modern world.[3] Pasolini's use of oxymorons, in a process he termed 'sineciosi' allows him to define Rome a 'miserable stupendous city,' or elevate an excavator to the symbol of a new proletarian civilization (Fortini, 1993). These are paradigmatic examples of 'translinguistic' activity which he theorizes in his writings on the relationship between language, poetry and experience. He explicitly rejects any idea of Saussurian arbitrariness, and avers that language – especially oral language – has archaic roots linking the speaking subject to a timeless language of reality.[4] Only poetry has the capacity to transform through language both experience and reality. In his article 'The written language of reality,' he advances this claim.

.... every poem is translinguistic. It is *action* 'placed' in a system of symbols, as in a vehicle, which becomes *action* once again in the addressee, while the symbols are nothing more than Pavlovian bells.

(Pasolini, 2005:198; original emphasis)

'The tears of the Excavator' is a poetic journey through Rome, from Trastevere to the borgate, to his own home in bourgeois Monteverde, revealing the city's oxymoronic totality. It is here that he uses the term 'stupenda e misera città' (stupendous miserable city), here that he talks about the passion of men 'allegri, inconsci, interi' (happy, unselfconscious, whole), here that he describes the Roman periphery as a 'meridionale periferia' (southern periphery), throughout expressing nostalgia for the marginality he had discovered in his days at Rebibbia.

> era il centro del mondo, com'era
> al centro della storia di mio amore
> per esso [...]
> [It was the center of the world, as it was
> At the center of my love for
> That place ...]

(Pasolini, 1976:97)

The poem has a bitter sweet character. The excavator can dig anywhere – in the depths of history, hopes and passion – but not in the pure forms of life that Pasolini recalls from his days spent in the borgata. The poem charts a journey starting from his innocent discovery of Rome, and ending with the cognizance that this pristine world will not survive in modern times. First published in 1956, 'The tears of the excavator' is the paradigm for conceptual operations pursued in all his Roman novels and films. These lines summarize the concept.

> [....]
> i soli africani, le piogge agitate
> che rendevano torrenti di fango
> le strade, gli autobus ai capolinea
> affondati nel loro angolo
> tra un'ultima striscia d'erba bianca
> e qualche acido, ardente immondezzaio...
> era il centro del mondo, com'era
> al centro della storia il mio amore
> per esso: e in questa
> maturità che per essere nascente
> era ancora amore, tutto era
> per divenire chiaro - era,
> chiaro! Quel borgo nudo al vento,

non romano, non meridionale,
non operaio, era la vita
nella sua luce più attuale:
vita, e luce della vita, piena
nel caos non ancora proletario,
come la vuole il rozzo giornale
della cellula, l'ultimo
sventolio del rotocalco: osso
dell'esistenza quotidiana,
pura, per essere fin troppo
prossima, assoluta per essere
fin troppo miseramente umana.
[...in the African sun, the restless rains
that turned the streets into muddy
torrents, the buses mired
at the end of the line in a corner
between a last strip of whitened grass
and some heap of rancid, fermenting garbage...
it was the center of the world,
as my love for it was the center
of history: and in this
maturity, still growing, there was
love all the same, and everything was
on the verge of becoming clear – it was
clear! That slum, naked in the winds,
not Roman, not southern,
not working class, was life
in its clearest light:
life, and light of life, full
in the chaos not yet proletarian,
as the rough newspaper of the
cell or the lasting waving
of magazines would have it: bone
of daily existence,
pure, because so
close, absolute because
all too miserably human.]

(Pasolini, 1976:97–98)[5]

We are still in the Neorealist tradition, modern but not postmodern. A measure of postmodern aesthetics can be taken from Adorno when he claims that for the sake of the beautiful there is no longer beauty (quoted in Bernstein, 2010:210). For Pasolini instead a new use of a classical aesthetic standard allows beauty to persist in modern times.

Spatial poetics

This discussion of poetics, place and meaning would be purely academic were it not for the presence some two hundred meters from

Gramsci's tomb of street art which manifests the precise principles we are reviewing here. Pasolini lives on in these works, showing the continued relevance of his poetic celebration of the Roman underclass. A primary standard of authenticity for street art is the specificity of place. Deracination is rejected in a strategy calibrated to resist the perceived destructive advances of global capitalism. Again, we are in Via Ostiense along whose course one finds the historic presence of Rome's industrial district, today almost entirely decommissioned. The narrative presented in these art pieces lays bare the rapacious character of global capitalism, revealing its propensity to extract value from the local community before moving on to the predation of new territories of conquest. This vision is fitly summed up in the graffiti piece entitled the 'Flight of capital,' painted by a pair of Argentinian street artists who sign their name as the Gaucholadres. This is part of a set of murals realized in 2013 on the walls of a grimy four-lane underpass which transits beneath an elevated rail corridor. Located some three hundred meters south of Gramsci's tomb, these murals infuse an abjected space with new meaning, achieving in pictorial form the translinguistic effect Pasolini realizes in his poetry. The murals reference the ravages of capitalism, and the promise of a new future achieved through resistance.

Discernable reference to Pasolini is exhibited in two mural portraits painted in the same underpass by the street artist Ozmo. One is of Gramsci, the other of Shelley (Fig. 4.3). Ozmo (2018) is one of Italy's most sophisticated street artists, and there can be little doubt that in these works he references Pasolini's elegy to Gramsci. To make the translinguistic point even clearer, Ozmo provides a third mural showing the image of the Tarot card representing Temperance. In Tarot, Temperance has the ability to harmonize opposites, symbolized by a hermaphrodite with one foot on land, one on water, holding two vessels one in either hand, with water flowing from the lower vessel to the higher one. The set of three murals is spatially organized in such a way that the viewer can see all three pieces by standing in a position half way down the eastern side of the underpass and looking west. Shelley symbolizes the classical standard of beauty, Gramsci the revolutionary force of resistance, and Temperance the capacity to harmonize this apparent opposition. The set of murals bears witness to the modern force of the past, tracing a genealogy from Shelley to Gramsci, from Pasolini to Ozmo, and on to the contemporary Roman public.

Ozmo is not alone in bridging the gap between Pasolini and the contemporary city. One of the most important murals in Rome adorns a former military barracks located right next to the underpass we just visited, in Via del Porto Fluviale, a name derived from proximity to the Tiber River. The former military establishment was squatted in the 1990s, and turned into a residence for about 200 homeless families.[6] Thanks to crowdfunding, and the work of Blu, one of the most celebrated street artist in Rome, a massive mural covers three of the building's walls, with an impressive linear extension of some 200 meters, three lofty stories high. Most impressive is the

Figure 4.3 Ozmo's rendering of Shelley. (Photograph taken in August 2020 © Gregory Smith.)

eastern wall with a scene of the apocalypse, a world being destroyed by fiery celestial incursions. In the middle of the mural is a tumultuous sea with an ark in the process of boarding the survivors of a biblical holocaust (Fig. 4.4). The ark flies the red flag of the communist movement, the same 'little red rag' Pasolini mentions in closing the poem 'The tears of the excavator.' A more direct link to the poem is the presence of towering excavators on the ark's main deck, which are busy hoisting the survivors on board. The yellow hue of these massive structures – the radiant color one expects to find in earth moving equipment – stands boldly out against the blue background. The positioning of excavators on a ship deck could hardly be a random choice, and undoubtedly references Pasolini's troubled yet salvific image.

Figure 4.4 Blu's eastern wall in Via del Porto Fluviale. (Photograph taken in August
2020 © Gregory Smith.)

Even the use of a biblical scene to illustrate a revolutionary message draws
from Pasolini's poetic language.

Pictorial deixis stands out boldly on the southern wall of Blu's Via del
Porto Fluviale mural, echoing in spatial terms the poetic technique we
saw in the a-Catholic cemetery. Most of the southern and western wall
is covered by zombie-like figures positioned literally inside the brains of
stylized human subjects, who are devoured by these sinister forces. In one
case an open skull is being filled with letters of the alphabet, in another
with rotting bananas (Fig. 4.5). These are the symbols of the discredited
world of advanced capitalism which deprives humanity of its dignity and
autonomy. One of the panels on the ground floor of the southern wall, on
an engaged pilaster at the center of the building, hosts the equivalent of

Figure 4.5 Detail of Blu's western wall. (Photograph taken in August 2020 © Gregory Smith.)

a cameo about two meters high. The panel boasts a realistically rendered image of the building on which the image is drawn, namely the building that stands before the viewer. The art piece shows what can clearly be identified as the neighborhood in which the building is located (Fig. 4.6). We can easily make out the Testaccio slaughterhouse, the Pyramid and the neighborhood stretching out along Via Ostiense, including the building standing before us. Just as 'Gramsci's ashes' links the reader to the exact experience of place, so this cameo reminds the viewer of the exact location of this artistic representation, brazenly underscoring the power of place.

There are many sources for this pictorial deixis, a story within a story which is the spatially specific account of the viewer assimilating the artwork. In Renaissance fresco painting a common feature is the use of an image painted on a wall representing the spectator's actual viewscape. The most famous example in Rome is in the Villa Farnesina in Trastevere, where a false perspective landscape portrays precisely what can be viewed

Figure 4.6 Detail of Blu's southern wall with an image of the Pyramid and the building itself. (Photograph taken in August 2020 © Gregory Smith.)

from the windows that pierce the frescoed wall. The story within a story fuses representation and experience, placing the viewer within the scene portrayed.

The last street art piece I would like to mention here is a mural done by Kid Acne, called 'Paint over the cracks.' The art piece is no more than the words contained in the title writ large (Fig. 4.7). It fittingly completes the rich set of art works that are scattered over the space of a few blocks, a poignant complement to the combat helicopter with which we started our tour. The helicopter is the symbol of a capitalist system which fragments the territory according to its own needs. The Kid Acne statement instead represents the action of local movements which reconstruct the territory's integrity.

Lefebvre (1991) noted long ago that capitalism fragments space according to its own needs. Resistance to fragmentation is orchestrated by citizens who attempt to maintain the unity of spatial experience, cultivating actions

Figure 4.7 Kid Acne's art piece. (Photograph taken in August 2020 © Gregory
         Smith.)

which preserve us from the crisis of presence. These pictorial and poetic
strategies create the robust sense of physical presence which is at the heart
of Pasolini's idea of the sacred. While Pasolini never defined the term, he
demonstrated that the sacred is a dynamic notion poised in the tension
between domination and resistance. The concept is fundamental not only in
Pasolini's oeuvre, but also in the dynamics which continue to regulate life
in the Roman periphery. One of Pasolini's most elegant expressions of this
tension is found in the film *La ricotta* (Pasolini, 1963), which I propose to
explore further after providing more conceptual background.

The sacred and the irrational

Pasolini's concept of the sacred follows a trajectory that starts with the
peasants of Friuli, moves through his passion for Rome's underclass, and
from the early 1960s onward projects its way to Africa, India and the global
south. A chief source of inspiration for Pasolini's thinking was Ernesto de
Martino, whom he applauds as the master of the Italian school of religious
studies. De Martino's most influential books are concerned with magical
practices in the Italian south and elsewhere. They include *Il mondo magico*
(1978 [1948]) and *Sud e magia* (1959).

I have noted the complexity of de Martino's intellectual orientation,
which includes interest in simultaneously promoting and discrediting
writers of the French ethnological school (Angelini, 1991:22). Indeed, de
Martino's academic production often contains manifest contradiction. One
of his key works, *Il mondo magico*, has been described as his most unortho-
dox Crocean work, marking a return to Croceanism during a period when
he was concerned about his isolation from the PCI (Cherchi and Cherchi,
1987:151–154). This is not unlike Pasolini's oscillation between idealism and

Marxism.[7] Notwithstanding abundant paradox, de Martino had the sizeable merit of bringing writings on religion to the attention of a relatively large reading public, especially when he joined forces with Cesare Pavese in launching the *Collana viola* as an Italian showcase for Europe's most important figures in ethnological and religious studies (Angelini, 1991:9–13).

Among the ethnologists promoted by de Martino was Lucien Lévy-Bruhl, who had significant influence on Pasolini (Barbato, 2010:258–259). Lévy-Bruhl is best remembered for his contention that 'primitives' have a completely different thought system from moderns, inasmuch as they deny the law of non-contradiction: '[t]he primitives show themselves insensitive to contradictions that we judge flagrant. [...] this indifference is one of the traits by which their mental habits contrast most visibly with our own' (Lévy-Bruhl, 1935:11). Pasolini extends this idea of absolute alterity from primitives to peasants and the urban underclass. This radical 'anthropological' divide fits perfectly with the essentialized spatial cleavage between center and periphery, combining social and spatial forces to generate a unique ontological context as a fundamental backdrop for Pasolini's narrative structures. The full expression of Lévy-Bruhl's radical alterity with its strong mystical vein is made manifest in Pasolini's works especially in the 1960s. The screenplay for the film *Teorema* (Pasolini, 1968), for instance, contains extensive hand-annotated reference to Lévy-Bruhl (Barbato, 2010:259).

In this period Pasolini also continues to be under the influence of Mircea Eliade, who extolls the mystical vein of human experience, and believes that religion as an irrational force can rescue humankind from existential void (Gasbarro, 2013). Pasolini dedicates the film *Medea* (Pasolini, 1969a) to Eliade thus underscoring an important source for his ideas on the sacred (Carbone, 2011:402). These various insights on the irrational and the sacred allow Pasolini to strengthen his use of the tension between positively-connoted mysticism and negatively-connoted bourgeois rationality as a central element in his artistic constructions. Yet Pasolini rejects Eliade's idea that the sacred can by itself shield against existential crisis. He believes that the sacred is an oppositional value changing according to shifting historical circumstances (Gasbarro, 2013:40).

An egregious example of peasant mysticism is found in the film *Teorema* (Pasolini, 1968) when the servant soars off into the sky, supported solely by the force of magical thought. This scene intends to demonstrate the limits of bourgeois rationality (Barbato, 2010:252). Yet shortly later Pasolini rejects the very idea of such rationality, seeing irrationality as permeating all levels of human experience. This is the idea contained in the poem 'Callas.'

> La tesi
> e l'antitesi convivono con la sintesi: ecco
> la vera trinità dell'uomo né prelogico né logico,
> ma reale. Sii, sii scienziato con le tue sintesi
> che ti fanno procedere (e progredire) nel tempo (che non c'è),

ma sii anche mistico curando democraticamente nel
medesimo tabernacolo, con sintesi, tesi e antitesi.

[The thesis
and the antithesis coexist with the synthesis:
this is the true trinity of man neither pre-logical nor logical
but real. Be scientific in your syntheses
which make you proceed (and progress) in time (which does not exist),
but be also mystical in applying a cure democratically
in the same tabernacle, with synthesis, thesis and antithesis.]

(Pasolini, 1996:738)

The issue is not the contest between the mystical and the rational, but the character of the 'real.' The real preserves the sacred while stripping away instrumental rationality to harmonize the individual and the environment, as we see in 'The tears of the excavator.'

[...] ad avere
il mondo davanti agli occhi e non
soltanto nel cuore, a capire
che pochi conoscono le passioni
in cui io sono vissuto:
che non mi sono fraterni, eppure sono
fratelli proprio nell'avere
passioni di uomini
che allegri, inconsci, interi
vivono di esperienze
ignote a me. Stupenda e misera
città che mi hai fatto fare
esperienza di quella vita
ignota: fino a farmi scoprire
ciò che, in ognuno, era il mondo.
[... to have
the world before my eyes
and not only in my heart, to understand
that few know the passions
in which I lived:
which are not fraternal, and yet
are brothers in having
the passions of men
who are happy, unselfconscious, whole
living experiences
I did not know. Stupendous and miserable
city which allowed me to have the
experience of life I did not
know: ultimately allowing me to discover
that which, in each of us, is the world.]

(Pasolini, 1976:93)

Unreflecting sharing in the hierophany of place gives reality its sacred character. This is the source of the happy-go-lucky attitude associated with life unencumbered by instrumental reason. Again, the attitude is best seen when the characters of *Street kids* burst out in song for no reason whatsoever, as seen with Riccetto.

'And now?' he said when the empty bus dropped him at Prenestino. He looked around, hiked up his pants, and seeing that there was really nothing for him to do there, he burst into song, philosophically.

(Pasolini, 2016:91)

Hierophany establishes a sacred condition which binds the world together in a seamless totality. Pasolini's graphic descriptions sacralize the person, just as the obsessive rendering of physical setting sacralizes place. Focus on the profound humanity of person and place is a technique he draws from Italy's distinguished artistic heritage. Pasolini had no background in filmmaking when he initiated his cinematic career, and drew inspiration for his visual language from Italy's religious painting, in particular from the Renaissance painter Masaccio. This is the ultimate source for the truth language of images. Following the technique used in these paintings, Pasolini's close-up facial images create a mood and tell a story. Masaccio's *The tribute money* renders the intense humanity of the scenes he portrays, an approach Pasolini imports into filmmaking (Gerard, 1983:45). Consider again the establishing shot of the film *Accattone* (Pasolini, 1961a), which sets the mood for the entire film. It is a close-up portrait revealing the enigma of the human face. *La ricotta* also dwells at length on portraiture, creating a mood, telling a story.

The civic sacred

A fundamental concept in de Martino's writing is 'presence' as the rational expression of human value in historically constituted society. Magic rescues the individual from the 'crisis of presence' which occurs when human suffering exceeds the limits of rational tolerability (de Martino, 1978). De Martino rejects entirely Lévy-Bruhl's characterization of magic as a kind of 'delirium' (de Martino, 1978:242). Magic instead expresses 'a dominant aim voluntarily pursued' in the effort to maintain a presence in society (de Martino, 1978:108). It is a socially grounded response to historically determined threats to human existence (Cherchi and Cherchi, 1987:117).

The social function of magic is the object of de Martino's classic study of the trance dance called the tarantella. In the Italian south, in Puglia, some women would routinely fall into a trance involving twitching and shaking caused, according to the folk belief de Martino documents, by the bite of a tarantula spider. De Martino's (1959) most important study of magic in the south dwells on this cult, offering various insights into this now famous practice. The ritual's most important feature is existential crisis, the risk of not being in the world. The crisis of presence has multiple causes, being partly a form of psychological dissociation, partly a form of alienation, partly a loss of subjectivity (Saunders, 1993:882). All of these features are associated with subaltern groups under the powerful exploitative control of a central force. Pasolini's familiarity with these southern cults is evident in a scene in *Street kids* where a woman falls into a kind of trance state.

> [Alduccio's mother] had jumped out of bed shrieking like a lunatic that she had seen the devil. She said a snake had come into the room, and was coiled at the foot of the bed, staring at her, forcing her to strip naked; and she had begun to shout. Then all day long, suddenly, she would start shrieking again [...]
>
> (Pasolini, 2016:179–180)

The idea that irrationality might have a rational social function is famously advanced in Émile Durkheim's *The elementary forms of religious life*.[8] As a social realist, Durkheim rejects the idea of a psychological source for social phenomena. Yet he still gives a psychological foundation to his realism, positing a rational purpose for the religious 'delirium' which ensures social order (Lukes, 1973:19). This is similar to the idea developed by de Martino, and had direct impact on Pasolini's theater piece *Orgia* (Pasolini, 2001), a gruesome story of sadomasochism practiced within a bourgeois household of northern Italy. In the program notes prepared for its 1968 performance in Turin, Pasolini states that he used Durkheim's ideas to create a theater space devoid of social conventions (Pasolini, 2001:320). There individuals can seek pleasure without the limitations imposed by membership of a social group. In Durkheim's study, the unfettered pursuit of individual pleasure leads to a condition of anomie, and eventually self-annihilation, which is indeed the outcome of the action enacted in the bourgeois bedroom. For Durkheim, and evidently for Pasolini, some element of community control is required to give desire a socially constructive outcome.

A fundamental element in neo-Durkheimian studies is the sacred as an element which stands above the individual, to wit society itself. This is the principle behind the notion of civil religion (Bellah, 1975), a concept that expresses and promotes the centrality of the community as the bedrock for individual experience. As we have seen, the importance of community solidarity in Roman parlance is captured in the dialect expression 'to act like an American,' used to sanction behavior when a person fails to support their congregation. While ostensibly a male concern, the capacity for concerted social action is also a recognized quality of women. For Pasolini, this solidarity is fundamental in human self-preservation, a form of community integration which ensures contentment without the anguish of individual consumerism (Sapelli, 2015).

In Pasolini's treatment we find multiple uses of the sacred: the quality attending to the individual, the community, the physical environment. These dimensions characterize spontaneous everyday life, fused together in profound, seamless experience. The sacred involves an unreflecting practice rather than a set of ideas (Gasbarro, 2013:39). In Pasolini's formulation the spontaneous unity of experience is a critical concept, and when the unity is disrupted so the crisis of presence ensues. De Martino sees this crisis as an affliction chiefly of peasant culture, while Pasolini understands it to be prevalent in advanced capitalist society. In that society the loss of subjectivity is determined by oppressive market forces which annihilate spontaneous social relationships, as well as free access to an understanding of the person, and the place where life unfolds. In real life there is no tension between rationality and irrationality, no hidden function supporting some higher unity. The sacred community is one where whole citizens pursue an unreflecting life which is happy and unselfconscious.

The modern crisis of presence stems from the domination of market logic over the spontaneous human condition. Of course, all communities are built on exchange, and in the vision supported by Pasolini true value is not what is transacted on the commodity market, but is instead expressed in a transaction carrying beyond the intrinsic worth of material tokens. This is the idea behind hierophany, a dimension bringing together the insignificant appraisal of the material object with value expressed on the higher plane of the sacred. Here is a description of value as understood by Eliade.

> Real value is only achieved when the social exchange is capable of overcoming the material quality of the signifier, the natural quality of the referent, and the savage solitude of the individual: this process of continual re-signification is the foundation of any civilization [...] because it guarantees humankind's presence in the world and in history.
> (Gasbarro, 2013:44)

Such an exchange wards off the 'entropy of civilization' entailed by the triumph of modern consumerist logic (Gasbarro, 2013:40). It expresses a historically specific condition and goes beyond that condition. It is witness to true human dignity manifest in peasant society and the urban underclass beyond the reach of modernity. It ensures untrammeled connection with archaic civilization, close to humankind's pristine origins.

In the 1960s Pasolini was not alone in ruminating on the meaning of the sacred. In those years Catholic circles were stimulated by a similar interest (Subini, 2006). The connection is compelling, given the fascination with Catholicism that traverses Pasolini's entire oeuvre, and which is given most obvious expression in his feature-length film based on the New Testament, *The Gospel according to Matthew* (Pasolini, 1966) (Golino, 2005:70). Inevitably, relations between Pasolini and the Vatican were far from idyllic, yet this film was hailed by the Vatican newspaper, the *L'Osservatore*

Romano, as being the greatest film ever made about Jesus (Osservatore Romano, July 23, 2014). The convergence of his reflections on the sacred and those of some Catholic writings is not coincidental. One source of affinity arises from background research Pasolini carried out in the early 1960s for his first film involving explicit biblical reference, which is *La ricotta*. On that occasion Pasolini spent several weeks in Assisi working with the progressive Catholic group, Pro Civitate Christiana, to understand how he could express his modern concept of the sacred through the vehicle of the New Testament (Subini, 2009:60).

The sacred character of the human physical condition has a long tradition in Catholic doctrine, and found unexpected support in Pasolini's writings against the legalization of abortion in the 1970s.[9] The importance of the human social condition was instead only given full expression at a later date, especially with the teachings of the Second Vatican Council in the early 1960s. We can summarize the evolution of a complex doctrinal issue by stating that the sacred in earlier Church teachings concerns mostly sacramental purity achieved individually through the services of the clergy. The progressive doctrines expressed at the Second Vatican Council shift the emphasis somewhat by giving more attention to the sacred congregation of the human community. This latter concern is most powerfully seen in John XXIIII's encyclical *Peace on earth*, which talks about the importance of a natural moral order standing above any possible legal mandate promoted by a given state system: 'laws and decrees passed in contravention of the moral order [...] can have no binding force in conscience' (Pope John XXIII, 2014 [1963]). This explicit invitation to civil disobedience is justified by the understanding that human society is divinely mandated, asserting that the sacred is a quality *in* rather than *beyond* the world. This formulation gives rise to progressive Catholic movements which join civic ideals to faith and commitment (Riccardi, 1997:144–146). It is also associated with Liberation Theology, uniting Catholics and Marxists (Gutiérrez, 1988). In those years there was an unexpected convergence, with Marxists like Pasolini writing about the sacred and progressive Catholics promoting civic goals going beyond sacramental compliance.

Although differently expressed in different periods, religion finds its way throughout the entire corpus of Pasolini's work, starting in the 1950s with the poem *La religione del mio tempo* (later published in Pasolini, 1961b). Potent religious references are prominent in the film *Porcile* (Pasolini, 1969b), where the protagonist wanders in the desert and trembles with joy at having killed his father and eaten his flesh in a grotesque parody of the Eucharist. Yet though concerned with religion, Pasolini's work is far from religious, as he repeatedly notes. In connection with *The Gospel according to Matthew* he was asked if he believed Christ was the son of God, to which he responded, 'Obviously not' (El Ghaoui, 2009:106). His idea of religion goes beyond religion in the ordinary sense, and is closer to the idea of the sacred.

He never defines the term, but reveals the meaning of the sacred in diverse settings. It would be impossible to pursue the full range of this meaning, but we can achieve a better understanding of the concept by examining a single setting in which the sacred plays a key role. An excellent setting is the film *La ricotta*. Given the importance of the concept in understanding peripheral urban life, and the fact that the film is set in Rome's periphery, I will explore his uses of the sacred in this connection at some length. The exploration will occupy the rest of the chapter.

Uses of the sacred in *La ricotta*

La ricotta (1963) was Pasolini's third experience in filmmaking and the first to make explicit reference to a biblical source given a radical interpretation. A refinement of his ongoing civic commitment, Pasolini's use of film was partly prompted by his conviction that in the 1960s images had begun to replace words as an expression of reality (Ward, 1995). Cinematic images are a 'natural language' which captures the sacred while reaching a broad public. His first films rely on crude filmic realism achieved by using film stock with intentionally coarse grain and high contrast, emulating documentary films of the time. Close-up shots privilege the human physiognomy, and nude figures with their sacred aura are on prominent display (Passannanti, 2019). Physical setting plays a key role in all of his early films. In *La ricotta* locations are brilliantly chosen in such a way as to portray the vision of sacred landscapes from Renaissance period painting alongside images of Rome's uncontrolled growth.

La ricotta is part of a collectively authored film entitled *Ro.Go.Pa.G.*, which includes other short films by the directors Roberto Rossellini, Ugo Gregoretti and Jean-Luc Godard. Although lasting only 35 minutes, Pasolini's contribution is a complete film in its own right (Subini, 2009:7). The story is straightforward. Stracci is the hero, an extra playing the good thief in a filmic representation of the crucifixion of Christ shot in the Roman periphery. The plot covers the activities of a single day. Evidently too poor to feed his family, Stracci goes without food and gives his bag lunch over to his wife and three kids. They consume the meal amidst Roman ruins in an open field near the filming location. Later in the day he comes into some money by selling the dog belonging to the film's diva. With the proceeds of the sale, he buys an abundant quantity of ricotta and bread. While feasting on this peasant fare in an isolated grotto, the film crew appears and serves him up a mocking banquet taken from the scene representing Christ's last supper. Shortly later, Stracci is hoisted up on the cross to shoot the crucifixion scene, where he truly dies of indigestion amidst the reproach of the assembled film crew, director and film producer. The message is clear. In the society of the spectacle, this true crucifixion becomes the ultimate gag for a voyeuristic bourgeoisie. The simple narrative is articulated within the structure of a story within a story, a modern interpretation of the mis en abyme. Most of the film is occupied by the efforts of the film crew to realize two tableaux vivants taken from important 16th century altarpieces.

The film is strongly personal, punctuated by polemical references to circumstances having nothing to do with the filmic action itself. For instance, the journalist who plays an important role in *La ricotta*, represented in a distinctly negative way, is given the name of the magistrate who had prosecuted the actor Franco Citti on trumped-up charges the year before as an attack against Pasolini with whom Citti was at the time filming *Mamma Roma* (Subini, 2009:128). Not surprisingly, *La ricotta* was seized by the same magistrate, and only released after various modifications, including a change in the journalist's name. According to some accounts, however, the real reason for seizing the film was because it seemed to represent the new-found alliance between Catholics and Marxist at the time when Italy's first center-left government was about to be formed (Subini, 2009:11).

The film consolidated Pasolini's public identity as a rebel and an iconoclast. It opens with a signed statement read aloud by Pasolini, in which he claims personal authorship for the film, saying he knows his reading of Scripture will generate scandal. Most scandalous is his assertion that the story of Christ is the greatest of all western narratives, a statement calculated to provoke communists who see the Bible as irrelevant, and the Catholic Church which strenuously defends its monopoly to interpret and represent scripture. Not only is the choice of material scandalous for a communist homosexual, so is his treatment of the subject matter. The statement read out in Pasolini's inimitable voice dispels any doubts about authorship, underscoring his poetic license in interpreting a key biblical scene. The effect he aims to achieve is a translinguistic modification of the uses of Scripture in the contemporary world. His idea that a poet can effect this transformation is set in polemical contrast to then fashionable structuralist ideas concerning the death of the author (Benedetti, 1998:11–12).

Produced in 1963, *La ricotta* represents an evolution in his approach to filmmaking, moving away from the purely aesthetic celebration of life in the periphery he adopted in *Accattone* (Pasolini, 1961a). *Mamma Roma* (Pasolini, 1962) and *La ricotta* both involve more critical commentary on underclass culture than found in his novels and his first film. The message of *La ricotta* is an almost pedagogical indictment of the sacred Pasolini associates with the Vatican. In some of his writings, like *La divina mimesi*, Pasolini speaks of a future without religion, asking if it is possible to imagine a church without the Vatican (Pasolini, 2008). This simultaneous rejection and embrace of religion hinges on a terminological distinction. The Vatican refers to the hierarchy of the Catholic Church, while the church concerns unreflecting peasant practices. This is an attack on Italy's official religion, although in the case that followed his indictment, Pasolini claimed that the religious framework was secondary to his exploration of the injustice to which the Roman underclass was exposed. He was initially convicted to a four-month suspended sentence later overruled by the court of appeals.[10]

As stated in *Poesie mondane* (Pasolini, 2012), Pasolini's artistic production is a force of the past, yet more modern than anything modern. He celebrates the classical tradition in many ways, including the use of Dante's terza rima in his poetry. In *La ricotta* he pursues this strategy by giving a modern interpretation to the classical idea of a locus amoenus or place of delight. The latter reference is key for understanding the film's complex layering of signification which juxtaposes classical and Renaissance ideals of beauty with the contemporary image of resistance. This analytical effect is achieved by combining historically distant semiotic readings of the landscape, and by recontextualizing Renaissance art pieces in a modern setting. From the outset, the life of Christ is presented as a true story supported by biblical quotations, and by visual reference to two of Pasolini's favorite masterpieces of Italian religious painting (Gerard, 1983:32–47). The enduring truth of the portrayal is also supported by filmic images emulating the Renaissance landscape used as a backdrop for religious painting in Italy.

The film has a serious intent, but makes abundant use of irony and humor. In one scene, Orson Welles – who impersonates Pasolini as the film director – states that the film is about his archaic Catholicism, a statement pronounced with an ironic smirk. There is a humorous quality to the scene where the actor playing Nicodemus picks his nose, forcing the director to interrupt the shot and start again. In another moment, actors echo in turn the call for a new scene; among the voices calling out, 'Make the other scene!' is that of a German shepherd dog that mouths the words in a humorous way. Stracci manages to sell the Diva's dog, and with these unexpected earnings races off to the ricotta stand in fast-forward motion with Chaplinesque music to match. On his way he passes a group of runners who have stopped along the roadside to perform a mocking portrayal of calisthenics. In the same fast-forward motion, he pauses briefly to give thanks at a shrine of the Blessed Virgin Mary. In another scene, the journalist passes carabinieri who are collecting flowers; in a statement that has no connection with the rest of the film they say they have nothing better to do.

Background elements are critical in conveying the message Pasolini wishes to transmit. Choices in musical accompaniment are coordinated with changes between scenes using color images and those shot in black and white, together guiding the viewer through the film. Locations take full advantage of Rome's rich visual potential, combining and juxtaposing images expressing contrasting value.

Painting and film

The central part of *La ricotta* is comprised of biblical scenes leading from the last supper to Calvary Hill. As we know, Pasolini was fascinated by the martyrdom of Christ, and in the following years developed his reflections further in the film *The Gospel according to Matthew* (Pasolini, 1966). The main theme in both films is the equivalence between the modern proletariat and the martyred Christ, where the suffering of Christ is identical to that experienced by the oppressed all over the world. Indeed, in Italy it is common usage to employ the word 'christ' to refer to any poor person who experiences suffering and oppression. The linkage between Gospel and contemporary injustice is underscored by the film's visual imagery drawing from a language well established in Italy's rich tradition of visual arts.

Italian filmmaking has a long tradition of borrowing from painting, the most distinguished examples being found in the films of Luchino Visconti (Del Guercio, 2007:55–60). Pasolini is part of this tradition, but his use of painting in film is inevitably distinctive and controversial. Most important is the way contamination and pastiche are deployed to elevate his narrative to a higher linguistic plane (Rhodes, 2007:56). Especially in his early films, inexperience in filmmaking is compensated by the use of painting as a source for his cinematic language. According to some claims, the reproductions of Masaccio used to inspire his cinematic portraiture were taken from black and white postcards which had the same coarse grain as his favorite film stock (Gerard, 1983). *La ricotta* combines the serene portrayals of pure humanity found in these early Renaissance paintings, with the bright hues and complex composition of paintings taken from a later period.

Interaction among contrasting images redefines our understanding of the world. It demonstrates the eternal return of the sacred, presenting Gospel as a paradigm for suffering in any historical period. The overarching framework of the story within a story draws from the long literary history of the mise en abyme, including Boccaccio's *Decameron* which is the source for a later film by Pasolini. *La ricotta* achieves profound semantic density by weaving into the narrative two fundamental altarpieces of Tuscan Mannerism which were little known to the general public at the time. These are altarpieces by Pontormo and Rosso Fiorentino, whose work was being reassessed in those years. Subini tells us that a recent volume on Pontormo, published under the inspiration of Pasolini's professor at the University of Bologna, Roberto Longhi, was on the set at the time *La ricotta* was being shot (Subini, 2009:149). His previous film, *Mamma Roma*, had indeed been dedicated to Roberto Longhi, acknowledging Pasolini's sensitivity to the visual imagery he had learned to appreciate under Longhi's teaching. The two Mannerist paintings chosen as the central images in *La ricotta* are particularly apt in building a story in a story, considering that for all their historical distance, these paintings are inspired by some of the same principles as Pasolini's modernism (Subini, 2009:152). The paintings structure the narrative in a way which encourages a multilayered reading of the film. One of the stories within the story is involved in comparing these two controversial painters of the early 16th century with the two controversial artists involved in making the film today. In *La ricotta* Orson Welles impersonates Pasolini, creating a pair of 20th century renegade artists who relive the revolutionary impetus of the two 16th century Florentine rebels. The philological sophistication of the choice is given by the fact that the two painters were suspected of being Waldensian heretics, those that put Christ above God the Father (Nigro, 2013:30–31). *La ricotta* does exactly that, implicitly claiming that Christ rather than being divine was an ordinary man.

Other details further the comparison suggested by the story within a story, like the liquid splendor of the colors Pontormo uses in *The transport of Christ* (1526 ca.). Nigro describes the colors as an explosion of 'desperate vitality' (Nigro, 2013:84). In *La ricotta* this vitality is contrasted with the desecrating attitude of the actors who realize the altarpiece in a tableau vivant, mirroring the Vatican's debased use of Scripture. The contrast between the splendor of the altarpiece and its cavalier representation is the film's strongest indictment of the Catholic Church. To signpost the film's narrative the scenes showing the representation of the altarpieces are shot in color, while the rest is shot in black and white, underscoring the contrast between the pretentious pomp of the Vatican and the sacred reality of everyday life.

Music adds another narrative layer, alternating between the sacred and the profane. On the one hand, we find music by Scarlatti and Gluck, accompanied by Tommaso da Celano's religious poem *Dies irae*. On the other, we find the *Eclisse twist* by Giovanni Fusco, the *Ricotta twist* by Carlo Rustichelli, and a band version of the aria 'Sempre libera' from Verdi's *La Traviata*. These are combined in often humorous ways, with the profane music being mistakenly played when the film enacts sacred moments. Interlocking visual and musical components ensure that the contrast between the debased sacred of the Catholic Church and the true sacred of ordinary suffering cannot be mistaken by even the most superficial viewer.

The film echoes semantic operations contained in Pontormo's masterpiece, *The transport of Christ*. To start with, Pontormo's composition is deliberately iconoclastic, citing while refuting Michelangelo's *Pietà*. Pontormo separates mother and child, and forces the viewer to engage in a complex relationship with the artwork. Michelangelo instead provides a canonic view of the mother and child in an artwork admired in a detached way as an objective portrayal of static reality. Pontormo exploits the physical layout of the chapel which still houses the altarpiece to create a story within a story. *The transport of Christ* is not a detached portrayal of a fixed historical scene, it is a story in the making, portrayed in an artwork so orchestrated that the viewer is bodily involved in the scene's unfolding. Ostensibly, *The transport of Christ* is a deposition involving all the canonic figures, yet the scene spills out from the canvas into the space occupied by the viewer. The technique is corporeal and deictic, referencing a dimension beyond pictorial language which is the spectator's own physical position. Christ is held lightly by two angels in the process of transporting his body to the altar located just below the painting. Christ is not only in the painting but is about to appear in the space occupied by the viewer, bringing the historical scene into contact with the immediacy of ordinary life.

The most remarkable feature of the altarpiece, however, is the compositional contrast between the void on which the painting is centered and the physical context of the surrounding chapel. *The transport of Christ* is housed in Florence in Brunelleschi's Capponi Chapel, in the church of Santa Felicita on the Arno's left bank not far from Ponte Vecchio. An engaging proxemic effect is created by placing the viewer between the pictorial scene of the transport of Christ's body, and the image of God the Father, once located behind the viewer at the entrance to the chapel (Nigro, 2013:71). While standing in the chapel, the viewer cannot fail to note that the eyes of the angels who bear Christ's body are directed outside the picture frame, contemplating the image of God the Father located behind the viewer. The pictorial integrity of the scene is secured by women in mourning who fix their gaze on the painting's central figure of the bereaved Virgin as they gather round her. The viewer is thus outside the image, but repeatedly drawn into it. One of the Marias holds in her hand a cloth with which to cleanse the wounds of the now absent body of Christ. This gesture contrasts with the cloth held by God the Father, a spiritual complement to the bodily ministrations shown in the canvas. Here body and spirit are mirrored, furnishing two different registers from which Christ can be apprehended by a viewer who is physically engaged in the scene's unfolding.

La ricotta condenses multiple perspectives on the filmic action, and deploys various techniques to engage the viewer in deciding how to read the film. One reading is simply the story of the New Testament set in the Roman periphery. The other is the passion of human suffering created by the indifference of modern capitalist society. The latter reading brings Christ outside the realm of iconic representation and into the company of ordinary humankind. The audience ostensibly can choose between these readings, but the latter emerges distinctly from the film's narrative fabric. Although lacking the architectural contrivances required to achieve the corporeal effect realized in the Capponi Chapel, *La ricotta* is so conceived as to compel the viewer to take an active role in appraising the scenes portrayed. No position is explicitly stated in the film, yet the critical interpretation is so obvious as to defy incomprehension. In the quotation shown at the opening of the film Matthew says 'He that has ears to hear, let him hear' (Matthew 11:15). The film evokes rather than declares. Like a biblical parable, *La ricotta* constructs meaning by combining elements known to the audience, structured according to the needs of an implied narrative statement. This truth is beyond words, and beyond the deceptive posturing of the Vatican. The film effectively urges the viewer to condemn the instrumental uses of the sacred and support the simplicity of the ordinary human condition.

Competing conceptions of the sacred are juxtaposed throughout the film. Images of the cross are particularly telling, whose use and musical accompaniment leave no doubt as to their true sacred value. Symbols of the passion are employed in a similar fashion. The crown of thorns in a solemn moment is shown suspended before an image of the Roman periphery, giving a sacred aura to what is masqueraded as a wasteland. The banquet table is set out sumptuously, and referenced throughout the film. It is a symbol of senseless opulence, implicitly contrasted with Stracci's proletarian feast of bread and cheese. As he is consuming this meal in an isolated grotto the diva appears and begins to laugh and mock Stracci, calling the modest feast the 'Stracci show.' The entire film crew gathers, bringing to the grotto the banquet table used in the film. They draw from this bounty to pelt Stracci with food in a mocking parody of the Last Supper, a perverse representation of Vatican charity (Cappabianca, 1998:71). The gaudy spectacle of bourgeois charity is contrasted with the dignified image of the marginal human condition.

Rosso Fiorentino's altarpiece *Descent from the Cross* (1521) instead operates on a different plane. Once more, vivid colors are so rendered in the film as to provide a dramatic contrast to the black and white scenes of everyday life. But more important is the stylized rendering of human figures. This is best seen in the unnatural inclination of the Virgin's head, a dramatically rendered expression of human suffering. Stylized visual representation is also seen in *La ricotta* in the fast forward scenes, or the speaking dog, or the improbable twist being danced on the film set, or the dramatic framing of Stracci's family shown from below as if they were divinities. *Accattone* had already inaugurated a filming technique defying the usual point of view editing that engages the viewer psychologically with the film (Rhodes, 2007:47). By distancing the viewer from the experience of viewing the film, the director invites the audience to consider their own perspective. Artistic representation is just that, a portrayal which is not the world itself but a particular reading of the world. This use of artistic liberty urges the audience to consider their own subjective response, suspending disbelief while remembering that point of view is a human construct.

Sacred and profane landscapes

In 15th century Renaissance painting the landscape began to acquire special narrative value. The landscape was 'humanized,' giving 'sustenance to the physical needs and spiritual yearnings of the men who inhabit it' (Turner, 1996:3). In the hands of such painters as Giovanni Bellini and Giorgione, landscape helps establish the mood of a painting. Light plays a special role, often contrasting the haze of the background with the crisp quality of foreground figures. Landscapes move from being an ornamental addition to being part of the narrative. The figures and the landscape fuse into a single narrative event, where the divine nature of humanity becomes such in a divine natural setting (Rosand, 1988:63–66).

Pasolini also uses sacred landscapes in his film, redefined according to his own poetic needs. Giorgione's *The tempest* uses lightning to create an intense mood, not unlike the thunder and lightning employed to create tension in *La ricotta* in the moments preceding Stracci's death on the cross. In the hazy background that serves as a backdrop to *La ricotta* we can perceive the pristine rolling hills of Giovanni Bellini's *Saint Francis in ecstasy*, with sparsely scattered buildings suggesting a classical landscape. The trees are those of Renaissance landscape painting, punctuated by the stone pines (*Pinus pinea*) typical of the Roman campagna. This landscape is given prominence through long and repeated panning shots, accompanied by solemn music. Juxtaposed against these canonic religious images are shots showing the circle of Rome's modern growth. The contemporary cityscape stands in symbolic representation of the uncontrolled development which produced the periphery, and with it Rome's underclass. The bucolic mood of Renaissance painting is rooted in the rediscovery of classical antique pastoral poetry, set in the hills and groves of rural Italy, far removed from the commerce and politics of the urban world (Rosand, 1988:67–73). This is the locus amoenus of the classical Arcadian poets, revived in the 16th century by Claude Lorrain to highlight connection between poetry and the visual landscape. *La ricotta*'s sophisticated editing turns the Roman borgata into an unexpected locus amoenus, a harsh reality home to humankind's true nature. This is the visual equivalent of the translinguistic action used in 'The tears of the excavator,' a hymn on modern human suffering.

The visual imagery sums up the idea of a stupendous miserable city, involving two aesthetic standards, one classical, the other contemporary. They are neither opposed nor mutually exclusive; they coincide. Historic Rome has been the setting for painting since time immemorial. Then with the massive and uncontrolled growth starting in the 19th century the city lost its artistic appeal, and Italian landscape painters fled from the city (Mammucari, 1990:8). Many of these painters became specialized in the Roman campagna that is still recognizable in *La ricotta*. Pasolini chooses a perfect location from which timeless Arcadia and the modern blight of uncontrolled growth can easily be juxtaposed. The location is today protected as the Regional Aqueduct Park, but in the days in which the film was shot the area was hotly contested. On the one side were developers, who wished to capitalize on the area's pristine open spaces. On the other, were urban activists who wished to preserve this as a green area showcasing the Roman campagna.[11] Landscape representations in *La ricotta* provide an eloquent case for the defense of this unique green space within the vast sprawl of the city.

The sacred in this world

Pontormo was a renegade from a religious standpoint: this is most evident in his now lost Last Judgment painted for the San Lorenzo chapel in Florence, where the figure of Christ was placed above the figure of God the Father. This was a heretical choice, attributable to the then fashionable Waldensian heresy (Nigro, 2013:30–31). To place a human figure at the center of religious attention challenges the canonical treatment of the sacred as being beyond the reach of the ordinary human condition. For Pontormo the sacred belongs to the community of the living. Pasolini chooses a similar orientation.

To make the point clear, *La ricotta* contains a contextualizing scene where a journalist – the emblem of bourgeois mediocrity – asks the film director four questions. These four questions and their relative answers establish Pasolini's conceptual orientation.

JOURNALIST: What do you wish to express with this new film?
DIRECTOR: My profound, intimate, archaic Catholicism.
JOURNALIST: What do you think of Italian society?
DIRECTOR: The most illiterate people and the most ignorant bourgeoisie of Europe.
JOURNALIST: What do you think of death?
DIRECTOR: As a Marxist it is not something I take into consideration.
JOURNALIST: What is your opinion of our great Federico Fellini?
DIRECTOR: He dances.

(PASOLINI, 1963)

The term 'profound, archaic Catholicism' references the idea of a church without the Vatican. Is it possible to cherish the sacred as expressed in everyday human life outside the political functions of the Vatican State? Pasolini's comment on the backward quality of the Italian bourgeoisie is an indictment of the complacency of his times, and the failure to live with a sense of civic commitment. Archaic Catholicism defends the centrality of the sacred and continues the film's opening statement about the importance of Christ in our western history. The this-worldly conception of the sacred is further reinforced by his statement that as a Marxist death is not something he takes into consideration. The final remark expresses Pasolini's rejection of the use of film as a pure aesthetic exercise, in favor of a politically engaged artform.

The director apostrophizes the journalist, calling him an average man, a colonialist, a slave-driver, a person without values. The film director pursues his Marxist interpretation, claiming that the journalist does not even exist. As a worker he is the mere instrument of capital, and exists only to the extent that capital wishes him to exist. The director of his newspaper and the film producer is one and the same person. If the journalist were to die on the set it would be excellent publicity for the film's release. The journalist, true to his bourgeois character, laughs foolishly, agreeing with the words but missing their deeper meaning. At that point, the film director turns his back on the journalist in a gesture that measures his inability to communicate with this pitiable expression of bourgeois mediocrity.[12]

Contemporary considerations

Another scene in the film shows an exchange between the cinematic Christ representing the Vatican and the real Christ represented by Stracci. As they are fastened to their respective crosses waiting to be hoisted up on Calvary Hill, Stracci turns to the filmic Christ to remark that he, Stracci, is so hungry he feels like swearing. The filmic Christ retorts jokingly that he will excommunicate Stracci for this profanation and then comments on the paradox that a deadbeat like Stracci should vote for the party of the owners, meaning the Christian Democrats. To this Stracci replies that the PCI is not much better; political parties are all the same. Stracci's statement voices Pasolini's usual critique of the PCI, and shows how remote the underclass is from the concerns of the communist party. Indeed, the underclass is entirely absent from political debate, which is one of the main points of the film.

Pope John XXIII, who is usually hailed as the first progressive pope in the postwar period, died in the same year *La ricotta* was produced. It is perhaps for this reason that the film was released to the public quite soon after being seized by the Italian authorities, and for this reason that Pasolini was treated lightly by the Italian justice system. This progressive orientation is seen once more today in the current pontiff, Pope Francis, one of whose first gestures was to receive in an audience the person most strongly associated with Liberation Theology, Gustavo Gutiérrez, granted after decades of isolation.[13] This shift may support the hope that the church can indeed exist without the Vatican, or within a Vatican which is closer to civic ideals. Perhaps the most enduring message conveyed by *La ricotta* is the aspiration that past and present, rational and irrational, the self and the other, can be reassembled in a way that represents genuine human value. Barbato puts it nicely when he speaks of a reaction against the desertifying consequence of this separation (Barbato, 2010:255). This idea was as radical in the 1960s as it is today.

Pasolini was a master narrator, conveying messages through poetry, novels and films, in addition to political and theoretical writings. His message is a compelling indictment of the ravages of uncontrolled economic growth. He advocates a humanistic conception of life rather than one founded on base consumer materialism. As I have argued, the idea of the sacred plays a central role in his narrative, a sacred which concerns the physical person, the community of which they are a part and the setting in which their lives unfold. From some standpoints we can see this as a conservative political outlook. Even the portrayal of the family as celebrated in *La ricotta* is a conservative institution: a procreative family comprising a man, a woman and their children. It represents a sacred enactment of the eternal cycle of human existence, appropriately set among the ruins of an ancient civilization.

His narrative deploys abundant spatial reference, and can legitimately be considered an environmental artform. He builds on simple elements used to explore complex issues and make sophisticated claims. Ordinary citizens do much the same in their narratives, although these constructions are not so sophisticated as those of an undisputed master. But sophisticated or not, geographically referenced citizen narratives reveal meaning which allows us to understand in more detail the way peripheral Romans experience their urban condition.

Notes

1 The imperative to reconcile past and present is clearly stated in 'Gramsci's ashes': 'Ma come io possiedo la storia, essa mi possiede; ne sono illuminato: ma a che serve la luce?' [But as I possess history, so it possesses me; I am illuminated by it: what purpose does the light serve?] (Pasolini, 1976:72).

2 Shelley's poem 'The mask of anarchy' is considered the first expression of pacifism, and was often quoted by Gandhi in his own political activism (Radhakrishnan, 2019): 'Stand ye calm and resolute, Like a forest close and mute.'

3 The other great British poet buried in the a-Catholic cemetery is John Keats whose 'Ode on a Grecian urn' celebrates the classical antique aesthetic standard, ending with the statement that beauty is truth. 'The tears of the excavator' functions as a modern transposition of this classical aesthetic pronouncement.

4 In *Heretical empiricism* Pasolini theorizes a distinction between spoken and spoken-written languages, where the former represent a direct relationship with nature, and the latter an instrumental and expressive relationship with work and society. For Pasolini this marks the distinction between Lévi-Strauss's prehistoric and historic phases of human society. He believes the ghost of spoken language persists in spoken-written language, revealing elements of an earlier civilization. These archaic elements are survivals in a continuous process of stratification (Pasolini, 2005:58–61).

5 Translated by Patrick Barron (Barron and Re, 2003:119–121).

6 https://www.repubblica.it/speciali/arte/gallerie/2014/11/25/foto/blu_murlaes_porto_fluviale-101387364/1/#1 (Accessed August 10, 2020).

7 In 'The tears of the excavator' Pasolini expresses this tension when he references his intellectual debt to the trio of Marx, Gramsci and Croce (1976:102).

8 First published in 1912, this work was translated into Italian in 1963 as *Le forme elementari della vita religiosa*.

9 Alberto Moravia, one of Pasolini's most intimate intellectual friends, remarked that Pasolini was a Catholic, provoking an incensed response by Pasolini (2008:105). The taunt was intended as an offence in connection with Pasolini's refusal to support women's right to abortion. For Pasolini procreation belonged to the sacred character of humankind. Yet on most positions, Pasolini differed sharply with organized religion, especially the Catholic Church.

10 The final sentence can be consulted at this site: https://www.olir.it/documenti/sentenza-07-marzo-1963-n-1020/ (Accessed August 12, 2020).

11 https://www.parcodegliacquedotti.it/storia-del-parco/ (Accessed August 2, 2020).

12 "Lei non ha capito niente perché lei è un uomo medio: un uomo medio è un mostro, un pericoloso delinquente, conformista, razzista, schiavista, qualunquista. Lei non esiste [...] Il capitale non considera esistente la manodopera se non quando serve la produzione [...] e il produttore del mio film è anche il padrone del suo giornale. [...] Addio." (Pasolini, 1963).

[You have not understood anything because you are an average man: an average man is a monster, a dangerous criminal, a conformist, a racist, a slave-driver, a person without values. You do not exist. ... Capital does not consider labor to exist unless it serves its own productive needs ... and the producer of my film is also the owner of your newspaper. ... Goodbye.]

13 Francis Rocca, 'Under Pope Francis, Liberation Theology comes of age,' *Catholic News Service* (September 13, 2013) http://www.catholicnews.com. (Accessed, August 15, 2020).

Bibliography

Agosti, Aldo (2012) *Storia del Partito comunista italiano: 1921-1991*. Bari: Laterza.

Angelini, Pietro (1991) 'Introduzione,' to Angelini Pietro (ed.) *Cesare Pavese e Ernesto De Martino, La Collana Viola: Lettere 1945-1950*. Turin: Bollati Boringhieri.

Asor Rosa, Alberto (1979 [1965]) *Scrittori e popolo. Il popolismo nella letteratura italiana contemporanea*. Rome: Savelli.

Banda, Alessandro (1990) 'Appunti sul leopardismo di P.P. Pasolini,' *Studi Novecenteschi*, 17:39:171–195.

Barbato, Alessandro (2010) *L'alternativa fantasma, Pasolini e Leiris. Percorsi antropologici*. Padua: Libreria Edizioni.

Barron, Patrick and Anna Re (eds.) (2003) *Italian environmental literature: an anthology*. New York: Italica Press.

Beck-Friis, Johan (1989) *Il cimitero acattolico di Roma: il cimitero degli artisti e dei poeti*. Rome: Grafica San Giovanni.

Bellah, Robert (1975) *The broken covenant: American civil religion in time of trial*. New York: Seabury Press.

Benedetti, Carla (1998) *Pasolini contro Calvino: per una letteratura impura*. Turin: Bollati Boringhieri.

Bernstein, Jay M. (2010) '"The demand for ugliness": Picasso's bodies,' in Bernstein, Jay M. and Thierry de Duve (eds.) *Art and aesthetics after Adorno*, pp. 210–248. Berkeley: The Townsend Center for the Humanities.

Cafritz, Robert, Lawrence Gowing and David Rosand (1998) *Places of delight: the pastoral landscape*. Washington, D.C.: The Phillips Collection.

Cappabianca, Alessandro (1998) *Il cinema e il sacro*. Genoa: Le Mani.

Carbone, Luca (2011) 'Pier Paolo Pasolini: tempi, corpi, mutamento sociale,' in Toscano, Mario Aldo (ed.) *Altre sociologie*, pp. 401–42. Milan: Franco Angeli.

Cherchi, Placido and Maria Cherchi (1987) *De Martino: dalla crisi della presenza alla comunità umana*. Naples: Liguori.

Del Guercio, Antonio (2007) 'Modelli pittorici in *Senso*: sentimento lombardo e composta naturalezza toscana,' in Galluzzi, Francesco (ed.) *Il cinema dei pittori. Le arti e il cinema italiano 1940-1980*, pp. 55–60. Milan: Skira Editori.

De Martino, Ernesto (1959) *Sud e magia*. Milan: Feltrinelli.

De Martino, Ernesto (1978 [*1948*]) *Il mondo magico: prolegomeni a una storia del magismo*. Turin: Boringhieri.

Durkheim, Émile (1963 [1912]). *Le forme elementari della vita religiosa*. Milan: Edizioni La Comunità.

El Ghaoui, Lisa (2009) 'L'enfer d'un monde sans religion: les derniers texts de Pasolini,' *Cahiers d'Etudes Italiennes*, 9:105–114.

Eliade, Mircea (2008 [1957]) *Il sacro e il profano*. Turin: Bollati Boringhieri.

Felice, Angela and Gian Paolo Gri (eds.) (2013) *Pasolini e l'interrogazione del sacro.* Venice: Marsilio Editori.

Foot, Paul (1981) *Red Shelley.* London: Sidgwick and Jackson.

Fortini, Franco (1993) *Attraverso Pasolini.* Turin: Einaudi.

Gasbarro, Nicola (2013) 'Sacralità come éthos del trascendimento,' in Felice, Angela and Gian Paolo Gri (eds.) *Pasolini e l'interrogazione del sacro,* pp. 39–53. Venice: Marsilio Editori.

Gerard, Fabien S. (1983) 'Ricordi figurative di Pasolini,' *Prospettiva,* 32:32–47.

Golino, Ezio (2005) *Tra lucciole e palazzo: il mito di Pasolini dentro la realtà.* Palermo: Sellerio Editore.

Gutiérrez, Gustavo (1988) *A theology of liberation: history, politics and salvation.* Maryknoll, NY: Orbis Books.

Iovino, Serenella (2016) *Ecocriticism and Italy: ecology, resistance, and liberation.* London: Bloomsbury Academic.

Lefebvre, Henri (1991) *The production of space.* Translated by Donald Nicholson-Smith. Oxford: Blackwell.

Leighton, Angela (1984) *Shelley and the sublime: an interpretation of the major poems.* Cambridge, UK: Cambridge University Press.

Lévy-Bruhl, Lucien (1935) *La mythologie primitive.* Paris: Librairie Alcan.

Lukes, Steven (1973) *Émile Durkheim, his life and work: a historical and critical study.* Harmondsworth: Penguin Books.

Mammucari, Renato (1990) *I 25 della campagna romana: pittura di paesaggio a Roma e nella sua campagna dall'ottocento ai primi del novecento.* Velletri: Edizioni Tra 8&9.

Maraini, Dacia (1998) *E tu chi eri? Interviste sull'infanzia.* Milan: Rizzoli.

Mariotti, Alessandro (2007) *Mircea Eliade. Vita e pensiero di un Maestro d'iniziazioni.* Rome: Castelvecchi Editore.

Moravia, Alberto (1929) *Gli indifferenti,* Milan: Edizioni Alpes.

Nigro, Salvatore Silvano (2013) *L'orologio di Pontormo: invenzione di un pittore manierista.* Milan: Bompiani.

Osservatore Romano, L' (2014) 'Il Vangelo secondo Matteo,' *L'Osservatore Romano,* 23 July.

Ozmo (2018) *Ozmo 1998-2018.* Paris: Crowdbooks.

Pasolini, Pier Paolo (dir.) (1961a) *Accattone.* Rome: Cino Del Duca.

Pasolini, Pier Paolo (1961b) *La religione del mio tempo.* Milan: Garzanti.

Pasolini, Pier Paolo (dir.) (1962) *Mamma Roma.* Rome: Arco Film.

Pasolini, Pier Paolo (dir.) (1963) *La ricotta.* Rome: Arco Film.

Pasolini, Pier Paolo (dir.) (1966) *Il Vangelo secondo Matteo.* Rome: Arco Film.

Pasolini, Pier Paolo (1976 [1957]) *Le ceneri di Gramsci.* Milan: Garzanti.

Pasolini, Pier Paolo (dir.) (1969a) *Medea.* Rome: San Marco Film.

Pasolini, Pier Paolo (dir.) (1969b) *Porcile.* Rome: Idi Cinematografica.

Pasolini, Pier Paolo (dir.) (1968) *Teorema.* Rome: Aetos Produzioni Cinematografiche.

Pasolini, Pier Paolo (1996) *Bestemmia. Tutte le poesie.* Edited by Graziella Chiarcossi and Walter Siti. Milan: Garzanti.

Pasolini, Pier Paolo (2001) *Teatro.* Edited by Walter Siti and Silvia de Laude. Milan: Garzanti.

Pasolini, Pier Paolo (2005) *Heretical empiricism.* Translated by Ben Lawton and Louise K. Barnett. Washington, D.C.: New Academia Publishing.

Pasolini, Pier Paolo (2008) *Scritti corsari.* Milan: Garzanti.

Pasolini, Pier Paolo (2012 [1964]) *Poesia in forma di rosa.* Milan: Garzanti.

Pasolini, Pier Paolo (2016) *Street kids*. Translated by Ann Goldstein. New York: Europa Editions.

Passannanti, Erminia (2019) *La nudità del sacro nei film di Pier Paolo Pasolini*. Salisbury: Brindin Press.

Pope John XXIII (2014 [1963])) *'Pacem in terris [Peace in the world],'* Rome: Libreria Editrice Vaticana.

Radhakrishnan, Sarvepalli (ed.) (2019) *Mahatma Gandhi: essays and reflections on his life and work*. London and New York: Routledge.

Rhodes, John David (2007) *Stupendous miserable city*. Minneapolis: University of Minnesota Press.

Riccardi, Andrea (1997) *Le politiche della chiesa*. Cinisello Balsamo: San Paolo.

Rosand, David (1988) 'Giorgione, Venice and the pastoral vision,' in Cafritz, Robert, Lawrence Gowing and David Rosand (eds.) *Places of delight: the pastoral landscape*, pp. 20–81. Washington, D.C.: The Phillips Collection.

Rossellini, Roberto (dir.) (1945) *Roma città aperta*. Rome: Excelsa Film.

Saunders, G.R. (1993) '"Critical ethnocentrism" and the ethnology of Ernesto De Martino,' *American Anthropologist, 95*:4:875–893.

Schneider, Jane (ed.) (1998) *Italy's 'southern question': orientalism in one country*. London and New York: Routledge.

Said, Edward W. (1985 [1978]) *Orientalism*. Harmondsworth: Penguin Books.

Sapelli, Giulio (2015) *Modernizzazione senza sviluppo: il capitalismo secondo Pasolini*. Milan: Mondadori.

Scollon, Ron and Suzie Wong Scollon (2003) *Discourses in place: language in the material world*. London and New York: Routledge.

Siciliano, Enzo (1979) *Vita di Pasolini*. Milan: Mondadori.

Subini, Tommaso (2006) 'Il dialogo tra Pier Paolo Pasolini e la Pro Civitate Christiana sulla sceneggiatura de "Il Vangelo secondo Matteo",' in Eugeni, Ruggero and Dario E. Viganò (eds.) *Attraverso lo schermo. Cinema e cultura cattolica in Italia*, vol. 2, pp. 223–237. Rome: EdS.

Subini, Tommaso (2009) *Pier Paolo Pasolini. La ricotta*. Turin: Lindau.

Turconi, Sergio (1977) *La poesia neorealista italiana*. Milan: Mursia Editore.

Turner, Richard A. (1996) *The vision of landscape in Renaissance Italy*. Princeton, New Jersey: Princeton University Press.

Verbaro, Caterina (2017) *Pasolini. Nel recinto del sacro*. Rome: Giulio Perrone Editore.

Ward, David (1995) *A poetics of resistance: narrative and the writings of Pier Paolo Pasolini*. Ontario: Associated University Presses.

5 Citizen narrative

Spatial self-representation in the Roman periphery

Background

We have explored at length the implications of what can be termed environmental narrative for our understanding of the city, generated by one of the acknowledged masters of Italian artistic achievement in poetry, literature and film. I would now like to turn to the geographically situated narrative of citizens residing in the periphery. In theory the modern commodification of spatial relations should stifle the integrated view of the city held by an ordinary citizen. Instead, this investigation demonstrates the persisting role of public space in supporting expressions of the self and the city in a hierophanous relationship not distant from that envisioned by Pasolini. The exercise described here elicited spatial narratives along with graphic representations of space in such a way as to document the idea of the urban territory that ordinary citizens carry around in their heads.

Space is a necessary ontological dimension of our existence, yet, as we have repeatedly noted, it has not received the scholarly interest it deserves. Asking for graphic representations of their urban experience is one way to compel citizens to express themselves in spatial terms. Physical space is important in our everyday lives, but this sensitivity is usually implied rather than stated, as there are few occasions on which a citizen is invited to externalize their spatial understandings. Wayfinding is the most important context in which environmental sensitivity is expressed, yet spatial experience has far wider ramifications than this.

Broadly speaking two competing metaphors describe how people find their way around, one involving complex structures, the other a complex process (Ingold, 2000:220). The first metaphor is assumed in cognitive psychology, and the complex structure is the mental map. According to this model, in order to understand how people orient themselves in space the citizen must be asked to externalize the mental map in graphic form. The other model, assumed in ecological psychology, instead sees map making as a specialized form of storytelling (Ingold, 2000:219). For many years I have operated mostly on the latter assumption in various explorations of geographical understanding (Smith, 2013). This is a hybrid approach, asking people to produce maps while telling a geographically situated story, in a process where the story is of paramount importance and the map is a pretext to keep people thinking in spatial terms. Map making in this ecological metaphor is a specialized branch of narrative understood as a central human proclivity (Ochs and Capps, 2001:2).

One of the most important scholars of cognitive mapping in urban studies is Kevin Lynch. His geographical sensitivity reaches maturity with the publication in 1960 of the volume *The image of the city,* but it has roots in explorations dating to the early 1950s (Lynch, 2002). Lynch follows the approach of cognitive psychology, and asks the subject to externalize the mental map in graphic form. He is especially concerned to identify features which express the citizen's relationship with the city in such a way as to chart its image. This is a paramount consideration in the assessment of contemporary urban quality, and many scholars have continued the effort to identify the physical features that make the city imageable. An excellent example of this is the work of Deni Ruggeri, especially where he calls attention to the emics and etics of imageability, namely the insider's view and the view from the outside (Ruggeri et al., 2018). Most work on imageability is concerned with its external features, delineating design factors which support citizens in the experience of the city. I am instead entirely concerned with the insider view: the way citizens respond to the physical environment without worrying about formal assessments of urban design. Indeed, poor urban design can generate rich narrative. Such is the case of the Roman periphery where parataxic aesthetics would cause most formal architectural appraisals to place these areas in the category of Junkspace. Any citizen of the periphery will recognize immediately that their environment does not meet the standard of Rome's city center. But in most circumstances they will identify meaning and find comfort in their poorly designed neighborhoods, partly as a reaction against the criteria of appraisal espoused by the central system. This is an aesthetics of resistance, not unlike the aesthetics expressed in Pasolini's work. In the terminology proposed here, we can say that these narratives express a dimension of the sacred which rejects the dominant value system and embraces the simplicity of ordinary life directly lived.

I have collected an abundance of maps in many Roman settings over many years.[1] Citizen-drawn cognitive maps following these procedures clearly belong to the genre of storytelling, and reveal an otherwise hidden dimension of urban experience. I can provide many examples of understandings revealed through this process. For instance, I was intrigued by maps collected in San Saba, Rome's first nationally funded public housing neighborhood, built according to a high standard of imageability. The maps I collected in 2010 offer two contrasting renderings of the neighborhood's central square, Piazza Bernini. This is a stately rectangular space framed by elegant early 20th-century buildings. In asking citizens to produce maps of the area, I expected them to recreate a rigid grid form, as indeed was generally the case. But in one instance a well-informed local resident chose to represent the square as a circle, and the surrounding road grid as a swirl of curved lines. My first instinct was to reject the representation as inaccurate, but then I realized that the map was not wrong, it was simply telling a different story. From the verbal narrative which accompanied the map-drawing exercise, it was clear that the swirls represented the reciprocal bonds of local solidarity, better expressed in curvilinear rather than grid form. Just as we have seen how Pasolini in some instances takes liberty with spatial relationships, so ordinary citizens do the same thing. Straight lines in our western tradition denote a principle of rationality with a hierarchically organized segmentation of society (Ingold, 2007). This is antithetical to an inclusionary principle which cuts across conventional lines of community segmentation - race, gender, age and class – with no necessary concern for formal design or political intention. From the conceptual standpoint, the circular map is perfectly accurate.

Cinematic images of Rome have profound impact on the vision of the city, both in the eyes of the visitor and its residents. Over the decades an abundance of films has been shot on location, heightening general interest in the chosen sites. Many of these locations are in the periphery, marginal areas with few visitors. This circumstance is such that residents of marginal areas made known through filmic portrayals are generally pleased when they see someone from outside, a visit which suggests familiarity with the recusant city image. I experienced this strikingly in an investigation of a classic Pasolini neighborhood, Tuscolano II, home to the public housing project featured in the film *Mamma Roma*. The episode says much about this method of investigating the periphery.

In asking citizens to produce maps the researcher must be alert to the various biases evident in the map-making process. Some are given by our western tradition, such as the primacy of cardinal north over south, or west over east. This is combined with our western tendency to privilege the top of a printed page over the bottom, and the left over the right, generating maps with north on top and west to the left (cf. Scollon and Scollon, 2003). These cartographic considerations are often combined with deictic interventions, where the interlocutor speaks with their hands, indexing physical relationships between their location and that of the physical features they wish to describe. The process establishes a relationship between language and location, fused together in a complex narrative gesture.

Citizens generally express discomfort if asked to draw a map since it is an unfamiliar narrative technique. Once they consent, there is inevitable reflection on how to produce a map that can enhance the story they are engaged in telling. For the map-making exercise all I provide the citizen with is a pencil and a blank sheet of A4 paper. The activity is generally carried out in some public space, like a coffeeshop, or a library, but sometimes at a person's home or at the university. The moment when the subject undertakes the creation of a map there is a transition phase of reflection and hesitation, when the resident must decide how the graphic image can explain the physical features we are attempting to explore. The subject will often rotate their body in physical space to gain inspiration as to how the map can mirror verbalized relationships, holding the blank sheet of paper in different orientations and positioning the sheet in the direction of the space that is to be represented. A key orientation choice is what should be at the top of the map. Once this is decided the drawer is likely to hover with the pencil over the sheet of paper, experimenting with alternative approaches before making the first mark. All of this is difficult to record, but shows that map making is part of a complex geographically-oriented narrative process.

I am providing these details to preface a map-making episode which involved the owner of the hardware store at the Tuscolano II public housing project (Fig. 5.1). I was with my group of Cornell students, and the resident was at special pains to present an accurate map, because he wished to impress on the foreign visitors the virtues of his neighborhood. After much turning in space, casting his gaze out toward the surrounding cityscape which was invisible from the enclosed vantage point of his shop, he suspended the pencil in the air for a moment, and began to draw. The first line was placed at the top of the map to signify its importance. The location described in geographical fact is not to the north, but to the south. This is the Roman aqueduct which runs just outside the project, and is featured in scenes of *La ricotta* seen from the south. The resident had organized the sheet of paper in portrait orientation and drew the aqueduct along the upper edge. The second line was perpendicular to the first running along the left-hand

Figure 5.1 Tuscolano II with its impressive modernist façade. (Photograph taken in August 2020 © Gregory Smith.)

side of the sheet of paper, connecting our physical location to the aqueduct. Having established this spatial relationship, the resident suspended his map making so he could explain in a verbal account the importance of this physical circumstance. He dwelled at length on the architectural excellence of Tuscolano II, making connections with the aqueduct to show that the public housing project was like the Roman monument in that both expressed the most enduring architectural standards of construction. He claimed that the central building – known as the boomerang in local parlance – was built with foundations 20 meters deep making it as solid as any Roman monument. He added that the boomerang had originally been encircled by an extensive paved area. He struggled to find the term for the paving material, and eventually, with gleaming eyes, averred that the surface material was cobblestone. To dispel any possible doubt, he added that this was the paving material used in ancient Rome. As it happens, the statement is historically inaccurate, but effectively conveyed the idea that we were not standing in the depressed periphery, and instead were in a part of monumental Rome. This is translinguistics at its best.

Narrative biases

Modan's (2007) sociolinguistic study of the politics of place in Washington, D.C., must rank as a supreme example of narrative analysis in urban research. Through a detailed ethnography of speaking she documents an extensive range of linguistic events which create and define place. Her narrative analysis looks at the intimate daily exchanges which objectify knowledge of the world, exploring the myriad influences and variations to which this knowledge is exposed. One of the characteristics of Modan's approach, however, is that its detailed analysis of verbal exchanges within intimate social environments cannot be applied to wider settings. Nor can it be structured to elicit specific features of the way the physical world is brought within the sphere of ordinary citizen perceptions and practices. In order to access a broader geographical area, and specify a narrow thematic focus, I devised the technique which is described below.

Narration involves the process of encoding a message and transmitting it to a recipient, and a critical analytical feature concerns the biases generated precisely by the perceived expectations of the target audience (Czarniawska, 2012:5). One of the potential pitfalls of conventional ethnography is the idealizing bias inserted into the relationship between the ethnographer and the ethnographic subject (Marcus, 1997). The ethnographer attempts to establish a bond of complicity with the ethnographic subject, becoming an insider capable of characterizing the community to an external public. This process creates a mise en scène which, according to Marcus, enacts and defends a neocolonial relationship between the ethnographer and the community studied. The process of ethnographic knowledge acquisition and ethnographic writing is of course complex, with much variation from one setting to the next (Clifford and Marcus, 1986). But here we are dealing with the stigmatized periphery of a major European capital city, in a setting with strong potential asymmetry between ethnographer and community, and the presence of a pervasive metanarrative framework concerning the periphery. The risk of a hidden bias is concrete. The perceived mission of social research in many parts of the world is to facilitate advances in quality of life, advances understood in terms of the grand narrative of western liberal decency (Marcus, 1997:90). In this complicity the ethnographer is easily seen as the advocate of neoliberal progress, and citizen narratives may well be biased to that expectation. Modan reduces biases by capturing events delivered in an unsolicited way. Narrative research based on interviews, such as that described by Landinin and Connelly (2000), risks unwittingly to encourage the ethnographic mise en scène described above. Is there a way out of this dilemma?

The way out proposed here is the organization of performance autoethnography (Denzin, 2018), an exercise where citizens from the periphery are invited to document their understandings of life in the periphery, and offer these reflections in a performance for other citizens from the same peripheral area. In the specific instance the setting for preparing the performance was a training program offered by the city of Rome, in circumstances I will describe in more detail below. My contribution to the program was an introduction to urban ethnography, which I structured as a seminar in which participants provided georeferenced accounts of life in the periphery, supported by various exercises. Thus, it was not a fact-finding exercise aiming to elicit denotative meaning, but rather a creative exercise in which participants were asked to operationalize their understanding of the periphery in a way that could reveal a deeper understanding of this setting. I was able in this way to create a mise en scène in which citizens produced information for other citizens, facilitated by an external researcher. The shared idea was to generate material which could be used in a public performance involving the seminar participants and directed at local residents.

The thus conceived activity stimulated reflection on the city and its contemporary uses, focusing in particular on the city's public spaces. The discussion was of immediate interest to anyone involved in the city, and was also intended to furnish viewpoints that could help shape the priorities of local administrators. The event was to be held in a community theater at Corviale, Rome's most stigmatized public housing project. This seemed to be a fitting location for an activity which wished to celebrate the city in all of its forms. Owing to various circumstances the date of the performance was postponed, and ultimately it was never carried off. The problem was getting non-professional citizen-actors to perform publicly. Yet, though never realized, the idea of the performance provided an alternative mise en scène to conventional ethnographies, and served to generate the material presented here.

The methodology was influenced by the writings of Norman Denzin (especially Denzin, 2010). The first point was to specify the aims and approaches of the research project. It should be noted that this was a self-selected group of keen supporters of urban life, all from the periphery and all hostile to the privatization of the city's public spaces. The city has come increasingly under the influence of non-spaces (Augé, 1995), especially evident in the growing importance of the quintessential non-space of the shopping mall, diminishing the scale and function of conventional public spaces (Cellamare, 2014). In any narrative there is a bias, and here the bias was the defense of the city as a shared space. In this experiment in autoethnography participants revealed their identity by sharing different ways of conveying geographically situated meaning in an urban environment. The exercise contained a denotative review of definitions of the city as a public space, but was chiefly a creative endeavor. Story writing can be like poetry in urban research. Rather than placing emphasis on conclusions, it attempts to enlarge our understanding of the world. It urges us to resist undemanding interpretations, and move closer to understanding what it means to be human in specific circumstances (Faulkner, 2009:16). This was the ultimate aim of the seminar.

The performance was intended to be a kind of 'mystory' described by Denzin, borrowing from ethnodrama, participatory theater, and the Brechtian learning plays that also influenced Pasolini (Denzin, 2010:58–59). Inevitably our approach differed in some ways from Denzin's design. The wall between the performers and audience did not entirely disappear, but the audience was expected to engage in activities which would bring them into the workings of the performance. Like the Greek chorus in Pasolini's *Orgia,* the public could comment on the action taking place and the stories told, and also generate their own stories and interpretations. Autoethnography allows citizens to insert themselves in accounts of the social world. Blurred genres are a fundamental feature in contemporary research, using multiple approaches to engage the public in a way that is dialogical and pedagogical, while at the same time placing the citizen at the center of the research process (Denzin, 2010:13). Denzin stresses the importance of the epiphany, the moment in which one realizes they have reached an understanding hitherto unexperienced. This is a critical moment in the process of living the world from the changing perspective of one's own evolving position. Preparing for the performance was a liberating exercise aiming to enhance citizen driven community action, while at the same time saying something about their understanding of the community. It was a way of furthering the process of a learning assemblage.

The setting and the exercise

The performance was scheduled for April 19, 2013, a 40-minute performance organized by a handful of citizen participants drawing freely from the texts produced in the workshop. This was to be the outcome of the seminar which met once a week and lasted from December 2012 to April 2013. I was contacted by the course organizer, and asked if I could direct a component of a training project which also involved other specialists in urban studies, as well as a handful of city administrators and one deputy of parliament. I was to provide a contribution that could address issues in urban ethnography. It was an exciting prospect, since a substantial group of motivated local citizens was expected to be involved over a period of several weeks. I was given complete freedom in organizing my component, and having been asked to grade student work I could monitor the quality of submissions. The student completer with the highest score in the course was awarded an important internship. Participants had been selected through a local competition, and the cost of organizing the project was covered by the local municipal authority. The screening of candidates led to the creation of a consistent group of motivated and capable local residents required to attend all the sessions and complete all the exercises.

Thirty-two citizens were involved in the seminar, fairly evenly divided between women (15) and men (17). Participant ages ranged from late twenties to mid-forties. While the initiative was sponsored by a left-wing administration, not all the participants embraced left-wing identity, although most clearly did. A couple of the participants indicated privately their support for a right-wing party, but they were also keen supporters of the classical idea of the city. In seminar proceedings effort was made to avoid situations which would split the group in political terms. The aspiration of the course was to provide all of the selected citizens with training in techniques which could be beneficial to a local administrator or urban activist irrespective of political orientation. From my standpoint this was an opportunity to provide community service favoring urban justice, while collecting a series of citizen narratives organized in a way dictated by the methodological needs of my narrative research.

The project was organized by one of Rome's 15 municipi.[2] This was the municipio known by the name of Arvalia, the only municipio in Rome to have a proper name in addition to the official number (XI). Citizens involved in the project came from Arvalia or neighboring municipi, generally part of the western corridor of the city which starts at the historic Aurelian walls in proximity of Porta Portese and Porta San Paolo, and extends all the way out to the city's far eastern border on the Tyrrhenian Sea. On the right bank of the Tiber, the corridor passes through the Magliana neighborhood, then borgata Trullo, and eventually encounters the public housing project of Corviale. On the left bank, it passes along Via Ostiense starting at the Pyramid and running well beyond Garbatella. Arvalia, like most municipi, has a radial form, extending from the consolidated historical center out to the far periphery. This configuration is part of an effort to integrate peripheral areas with the center. The empowerment of municipi started in the 1990s, when they were given modest budgetary autonomy, used to support expenses such as child care, or recreational facilities for the elderly. The budget can also be used for other services, like the training program described here. Municipi are large territorial entities with substantial populations. Arvalia, for instance, has a population of some 155,000 residents.

I organized the seminar as a sequence of activities, starting with basic considerations like the graphic representation of space, and from there progressed to the creative use of location in citizen narrative. These activities were all focused on the participants' place of residence. The structure of the single submissions was loosely organized, and the most complex submissions, which were the final narratives, could draw inspiration from any source, including literature, television, film or comics. The two hundred pages of texts described here were produced by 32 participants over a period of five months. All participants were native Italians and wrote in their native language. Each activity was discussed with the group before it was undertaken, after which participants were given two or three weeks to produce their submissions. The initial texts were individual exercises, while the final narratives, the most complex and perhaps most interesting, documents, were group products involving five subsets of students based on residence within more or less homogeneous geographical areas. This is the sequence of events prescribed in the workshop.

> Individual exercises: Draw a neighborhood map. Comment on the map-drawing effort. Describe the emotional experience of drawing the map. Define the concept of the public in an urban environment, and assess your neighborhood in terms of this idea. Produce an itinerary in which you guide an imagined visitor in an intimate local tour.
> Group exercises: Produce semi-fictional stories using any narrative technique. Compile the results for the performance. Enact a public performance followed by a debate.

The first step in the series was to encourage participants to think spatially by creating a graphic representation of their neighborhood. In order to prepare participants for this exercise I discussed other citizen maps I have collected over the years. I also provided an overview of Kevin Lynch's work to show the relevance of cognitive mapping for urban studies, and furnished background on the map-making process. A particularly useful starting point for this kind of exercise is discussion of the five elements which Lynch (1960) claims describe the city image: paths, edges, districts, nodes and landmarks.

In the first exercise, participants used an A4 sheet to draw a simple line image representing their neighborhood. The seminar was organized in Italian, and the word used for neighborhood was quartiere. Words and images were combined by having participants accompany the map with a written description of the map-drawing activity. As a separate exercise, participants were asked to comment on the experience of drawing the map. The drawing exercise was part of a composite visual text, bearing in mind the weak separation in informational terms between linguistic and plastic representations of visual experience (Bal, 2003). Just as some literary texts only make sense visually, the inexpert maps produce here only make sense if the drawer illustrates in words the meaning they intend to convey. Map quality is of no importance, and maps simply provided support for verbal narrative. Starting the seminar with a map-making activity helped promote discussion on the importance of physical geography in structuring experience. We need to remember that maps are never neutral or objective. Researchers and citizens alike must cultivate sensitivity for the implicit discourses that go into map making. In the seminar setting, map production is a conscious part of a learning assemblage, ultimately restructuring socio-spatial relations which receive no support in the formal system of shared representation (McFarlane, 2011:27). The effort of the seminar was precisely to give voice to spatial awareness, to illustrate the character of spatial signification in a discredited urban environment.

This was the first of a five-part set of exercises. The second exercise was denotational, defining the idea of the 'public' in individual submissions. Unsurprisingly, the results of this exercise were convergent, all stressing the public sphere of the city as a primary dimension of human experience. The third exercise explored the meaning of selected physical locations in the neighborhood as semantic components of a spatial lexicon. In this exercise participants created itineraries which involved key elements in defining personal identity, specifying in the process places which constitute the lexicon of spatial language. The fourth exercise was a group activity in which participants produced short stories highlighting the spatial features which give poignancy to their narratives. This activity explored the evocative capacity of place. The final part was to be the performance, followed by a debate on the city as a shared plural space, stimulating reflection on the role of place in defining the community. For contingent reasons the performance was never staged, but constituted the mis en scène for the realization of a rich set of maps and texts.

Maps

The city of Rome is divided into multiple administrative and topographical units. The historic system involves concentric divisions starting with the rioni at the center, moving to quartieri occupying the first ring of 19th-century growth, from there out to the suburban districts, and finally the outer ring of the Roman campagna. As we know, the latter districts were filled with building sprawl dating mostly to the mid-20th century. Since 1961 the Roman campagna has been divided into zones known as the Agro romano, of which today there are 53.[3]

The municipio of Arvalia contains two quartieri (Portuense and Gianicolensi), two suburbs (an extension of the two quartieri, using the same names) and two Agro romano zones (Magliana Vecchia and Ponte Galleria). The other way of classifying the city is by urbanistic zone, established in 1977.[4] These were part of a new effort by the city of Rome to track urban growth. Arvalia is comprised of seven urbanistic zones, of which Corviale is the most famous owing to the presence of the public housing project having the same name. Of all these units only the municipio has an administrative function, with elected officials and a budget. All the other units are important chiefly from the statistical standpoint. Citizens may or may not use any of the boundaries defined by these units in organizing their perceptions of place, constituting as they do formal delimitations which generally have little relevance in everyday life. [5]

In the seminar described here, not all participants came from Arvalia. Of those outside Arvalia, one came from the other side of the Tiber River (Ostiense), another from a neighborhood north of Arvalia (Massimina) and one from a small town to the north of Rome. A fourth came from Tor de' Cenci located to the south of Arvalia. Although the urbanistic zone of Corviale is found within the territory of Arvalia, the housing project bearing this name belongs to the neighboring municipio. We had one participant from this municipio. As we have seen, Ostiense is the site of Rome's early 20th century industrial district, today much gentrified owing in part to the presence of one of the city's state universities. Massimina is a largely self-built area falling outside the ring road. Tor de' Cenci is an Agro romano zone with abundant recent construction, part of Municipio IX.

Within Arvalia the participants came from these areas: Marconi, Magliana, Villa Bonelli, Portuense and Ponte Galleria. The area of Corviale was partially included. Marconi was developed as part of the 1931 plan. It starts on the outskirts of Trastevere before it crosses the Tiber and joins on the other side of the river Via Ostiense and other roads leading out to the coast. It is today a dense residential area. As we know, Magliana is located on the right bank of the Tiber, and is one of the most densely built parts of the city. Villa Bonelli is nestled in the hills to the west of Magliana, separated by Via della Magliana and a railroad. It is an attractive neighborhood characterized by developer-built homes dating mostly to the 1970s. Portuense is a mixed area, with some spontaneous postwar growth, but also much attractive developer-built real estate. A consular road gives its name to Portuense, running from the historic walls of the city out to the port of Rome. In proximity of this road we find Corviale, with its housing project surrounded by a wide variety of different building types, including self-built homes from the mid-20th century, and high-income developer-built housing from the early 2000s. Ponte Galleria was founded in the 1920s when reclamation works were carried out by the fascist government. It once hosted industries, and is today known for the presence of Rome's facility for detaining undocumented foreigners. This western corridor of the city is characterized by a variety which is emblematic of Rome's checkerboard growth, reflecting alternating historical periods each with its own trajectory.

Corviale is sandwiched between two of the city's most extensive parks, Valle dei Casali and Parco dei Massimi, both managed by a regional authority called Rome Nature. Combined, they create an enormous green wedge extending along the city's western corridor, contained within the Great Ring Road (GRA). Just outside the GRA is Rome's landfill, Malagrotta, the largest landfill in the world when it was closed in 2013, thanks in part to intense environmental activism which continues on this western fringe of the city.

This rich variety of cityscapes gives rise to a rich variety of maps. The maps exhibit different levels of draftsmanship, and different orientations. As we have seen, north is not always at the top of citizen maps, which are instead commonly oriented to a local landmark or a geographical feature, like the Tiber River. This is seen, for instance, in the map of Ostiense (Fig. 5.2), one

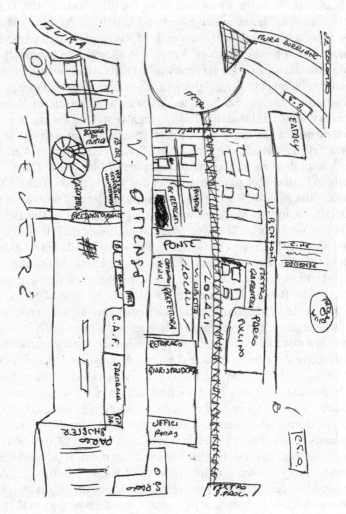

Figure 5.2 Citizen map of Ostiense collected in 2013.

of the more successful in terms of graphic rendering. This map allows the Tiber River to define a visual frame which also includes the Pyramid and the a-Catholic cemetery.

A feature of narrative relevance is the point of departure for realizing the map. Most participants had a sense of the extension of their quartiere, and in about a third of the cases, including Ostiense, the residents started their map with the neighborhood boundaries. But boundaries are an abstraction, and require careful thought before they can be established, especially since most quartieri exist in local perception rather than as a formal territorial unit. A common cartographic starting point is a feature linked to some everyday experience, especially the home which constitutes the starting point in another third of the maps. In other cases, the starting point is some prominent physical feature, especially the main square, if such exists, or a main road which cuts through the neighborhood, allowing access to the area from an external location. Like the boundaries of the neighborhood, having a main road suggests the presence of a discrete territorial unit with a separation between an 'inside' and an 'outside.' Monuments can also define the territory, and furnish a starting point for a map. All of these considerations were discussed at length with the group, reviewing the range of information expressed in maps understood as narrative elements. These maps reveal the importance of place as expressed through detailed knowledge of the local environment.

The experience of map making

Rome's variegated urban environment constitutes an effective grammar replete with potential for creative expression, affording the citizen an array of semantic tools which can be selected and combined in myriad ways to convey a wide range of intimate personal experiences. Cartographic vision is an assembly of heterogenous elements more or less consciously combined. Rarely are citizens called upon to externalize this knowledge in a systematic way; the map-making process unveils the mystique of place viewed through personal style and choice. Laying bear this intimate knowledge can be an intimidating experience, stripping away the cloak which shields intimacy from public disclosure. Most initial map-makers express discomfort with making a map. This is the description of the experience provided by the author of the Ostiense map.

> It is the first time I have drawn a map of my neighborhood. Initially I thought it was a trivial exercise. But then as I proceeded, I realized that I was not simply inscribing a real neighborhood black on white; instead I was putting into relief what is for me an indispensable world.
> (Male, Ostiense)

The reflections shared by participants reveal that urban geography is not a straightforward expression of knowledge, but rather a creative construction fusing together various elements into a kind of unity. The citizen from Ostiense notes that the neighborhood is not a 'real' space, the same for all citizens, but rather a source of meaning potential the individual must synthesize through a creative process. The idea of spatial meaning potential is nicely expressed in the following statement written by a woman participant from Magliana, summing up a poetics of space.

> A combination of lines that trace corners, curves, circles, that join and delimit spaces, that open on other lines, corners and curves. Intertwining, colliding, intersecting, travelling parallel while never touching, or constantly changing direction thus entering into collision with other shapes and forms. The lines fuse with our lives; lines that guide our unfolding existence. These are the places and pathways along which our life choices travel, choices which are conditioned by the contours of the landscape. Such is a map, a map of our city, of our neighborhood, of our lives.
>
> (Female, Magliana)

These statements have an autobiographical character, revealing something about the relationship between inner motivations, emotional states and the physical world in which lives unfold. Within the assemblage of elements one inevitably finds perceived notions about the social constitution of the world implied by the built environment. We see this in the following statement written by a resident who uses the English word 'privacy' – for which there is no effective Italian term – to illustrate her thoughts about what she perceives to be a less than ideal physical world. According to her view, private space is vital to personal well-being.

> In the little street where I live the space between one building and the next is almost nonexistent, leaving no room for privacy (original Italian = 'privacy'). By just looking out the window you unintentionally witness the lives around you. Lots of suicides take place here.
>
> (Female, Marconi)

Cities always contain tension between public and private, the public sphere being where one finds symbols of shared identity. Rome, of course, has an abundance of public symbols with deep historic roots. Peripheral neighborhoods certainly contain physical features with landmark potential, but none on the scale, say, of the Colosseum. This may seem irrelevant in the periphery, but Rome is a total urban experience, and when considering one's own neighborhood an implicit standard of comparison is provided by some of the most noted monuments of western tradition. By the same token, although overwhelmed by monumental density, even in central Rome intimate physical features have potential to become landmarks, like a street corner, or an ice cream parlor, or a park bench. Inevitably the periphery pales by comparison to the center in terms of landmark recognition. But this is not necessarily a negative consideration, and indeed discussion of apparently bland local monuments can heighten the sense of the value of the neighborhood, where local knowledge and solidarity prevail over obvious monumentality. In this case, the absence of external recognition gives legitimacy to intimate local practices which express resistance to the stigmatizing metanarrative. This reflects the essence of community life, where landmarks are created through the actions of ordinary citizens whose thoughts and actions endow place with distinction. Here we have a reflection on monumental features of the periphery which stems from the map-making activity.

> The point of departure for drawing the map, after many false starts and erasures, was the service station I see from my bedroom. After having tried various departure points, I settled for this site because that is where each day begins and ends.
>
> (Female, Portuense)

Wayfinding in human experience has little to do with cartography in a formal sense. Wayfinding instead concerns memory and repeated practices constituting physical space as an 'objective' realm in which citizens express their lives (Ingold, 2000). It is memory in physical form, a set of signposts which elicit experiences anchored in space. Spatial features have a subliminal dimension – a quodlibet that goes beyond verbalized experience – with powerful evocative capacity because these features solidify a dense array of images, memories and meanings. The hand-drawn map is a part of a visual narrative which stimulates the effort to find words for nonverbal cues. Spatial orientation is fundamental in story-telling: it is not just a setting but a story in its own right.

> Drawing a map means to discover oneself: it brings to mind all those memories, all those journeys I used to undertake; the map is thus an expression of oneself.
>
> (Female, Tor de' Cenci)

An urban space is a living canvas on which different narratives can be inscribed. Yet it is not a blank slate, a realm of infinite plasticity capable of acquiring any signification. It represents the complex intersection between objectivity and subjectivity, between the visual and the textual where one dimension extends the capabilities of the other. Nor is it the naturalized space or truth language theorized by Pasolini, although iteration in narrative gives spatial features an appearance of fixity. There is a porous relationship between people and places where one shapes the other, in an aesthetic as well as a normative process. The fact that these areas are often parataxic may appear to be an aesthetic impediment, but parataxis also endows space with a flexibility separate from the intentions of the builders of space, leaving wide margins for innovative interpretations. Normative features may change from one speaker to another, from one moment to the next. Spaces in perception may be too cramped, as in the account from Marconi given above, or too diffuse to create any community sense. These normative considerations, folk norms given local specificity, call to mind Jane Jacobs's (2016) discussion of the role of density in determining the quality of urban life.

Some spatial circumstances may be construed as discrediting features even in autobiographical narrative. In these cases, a backshadowing narrative technique is often involved, using spatial discredit to give dramatic prominence to positive outcomes. The most common technique is to show how the discrediting feature prompted a virtuous response, allowing the citizen to overcome the construed adversity of place. In common Italian perception, cities are associated with an ideal organizing principle capable of uplifting society in moral terms. With Rome's unregulated peripheral growth, this ideal ordering principle often lacks physical expression, and the penumbra between expectation and experience furnishes an excellent setting for backshadowing narrative. A hostile urban environment is a sinister element and an excellent setting for a film noir.

> I never imagined that I would have to describe the place I live in with a map: it was a dramatic experience because it made me realize that I live in a place which is to say the least absurd.
>
> (Female, Ponte Galleria)

Definitions and assessments of the public sphere

The public sphere in the classical western tradition is a critical urban feature. Studies of well-conceived urban spaces have an important modern history, including Camille Sitte's (1979) wonderful diagrams that capture the multiplicity of figure ground relationships generated by the European city-building tradition. Italy has a history of excellence in urban design seen from the standpoint of figure ground expectations, with particularly egregious examples found in the Italian middle ages (Piccinato, 1993). The glories of these cities are anchored in their public spaces considered as primary urban elements (Rossi, 1984:86). The political corollary of excellence in urban design is a compelling sense of community membership, the foundation of the 'right to the city' (Lefebvre, 2006).

It is paradoxical that a defining characteristic of the Roman periphery as an urban space is precisely the general absence of effectively designed public spaces, yet a strong sense of attachment to the city as a political project. Indeed, there seems to be an inverse relationship between political sensitivity to the city and its physical expression in good design. Roman citizens have an informed awareness of the physical features which characterize a well-designed city, and their absence seems to stimulate a strong sense of local attachment. The disjunction between expectation and experience can generate ironic narrative, but also serious effort to strengthen the practice of the city. For the seminar participants the idea of the public sphere was of central importance because it is what the city is all about, representing the highest aspiration of western civilization. All workshop participants accepted the hierarchy between public and private as a governing framework in which urban narrative is articulated. Consider this definition of public space offered by a resident of the public housing project in Corviale (the citizen's map is shown in Fig. 5.3).

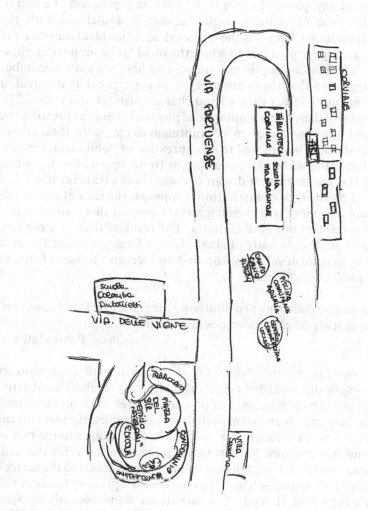

Figure 5.3 Citizen map of Corviale collected in 2013.

Public space is a place that belongs to all individuals with no distinction of any type. It is in this space that single individuals become a community, and for this reason that no individual can exert power over any other. In this space individuals become and act as a community, while in a private space the individual maintains independent identity and acts only for the self.

(Male, Corviale)

This is a textbook definition of the public sphere in western political discourse, and could easily have been taken from the fundamental principles of the Italian constitution. Formal equality and shared political commitments allow the individual to perfect their human condition.

As in hierophany, physical space is the profane expression of a sacred ideal which is the community itself, intimately associated with the quartiere. Interaction between the physical and the ideal elevates the individual to a higher plane, even when the ideal lacks support in physical design. The relationship is not static, and here we must remember Latour's suggestion that the community is as conceptual as it is real, a process of becoming rather than a fixed characteristic (Latour, 2007:27). While no one will deny the advantages of physical spaces expressing the canonic standards of good design, commitment to the public ideal allows any physical space to be elevated to an expression of public attachments. In cities all over the world there is a retreat from support for the public sphere, and this has important design consequences (Trancik, 1986). Yet notwithstanding the lack of institutional support, the idea of public life in the Roman periphery is very much alive. Consider this statement formulated by a resident of Ponte Galleria. The resident lives in a modern neighborhood made up mostly of single-family homes, designed almost exclusively by private developers who had no incentive to build features expressing public attachments.

It is not easy to define my neighborhood; or perhaps it is too easy, considering that my neighborhood does not exist.

(Female, Ponte Galleria)

The real message is that the community should exist both as an abstract ideal and a physically grounded neighborhood. Many neighborhood activities are structured by the presence of services tailored to the needs of residents. Yet neighborhoods are not expected to be self-sufficient, since citizens have access to a much broader area, evident in the linkages to the rest of the city found in narratives. We saw in accounts from the 1950s that citizens from underprivileged neighborhoods have liberal access to the center's most protected public spaces. Thus, the quartiere is a primary frame of reference satisfying a host of needs, but citizens are simultaneously engaged with the wider metropolitan area. Neighborhoods typically offer the full set of retail and service activities that citizens require in daily life, from food to clothing, to cleaning services, to forms of entertainment. An important neighborhood function is the provision of cultural services which the narratives collected here deem inadequate. Such activities include music, theater, lectures and debates. Naturally, the presence of specifically defined public spaces would facilitate these activities. Community services, such as health care and public administration, are also expected. This is the practical side of the neighborhood. The ideal side is the civic culture visible in contemporary debate as it is in the narrative creations of ordinary citizens – it is a value in its own right.

We have noted how communities are built on exchanges, and Eliade's idea of an exchange carrying beyond the intrinsic value of material tokens. This idea is present in the thinking of most citizens who recognize that some local facilities combine a strong focus on civic culture with the provision of important services. Local open-air markets providing retail mostly of fresh food products are one such facility. In Italy these markets promote community identity and encourage locally based solidarity (Black, 2012). There are about 130 such markets in Rome, organized by the city's department of commerce. They satisfy more than commercial needs, and represent a hub around which a range of retail activities is organized. Sharing in the life of the market is a way of sharing in the community, satisfying material needs while coming into contact with other local residents (Fig. 5.4). It is for this

Figure 5.4 Neighborhood market at Tuscolano III. (Photograph taken in August 2020 © Gregory Smith.)

reason that some of the most charming narratives produced by citizens are set precisely in the market.

Two other facilities which combine support for civic culture with the provision of a vital service are public schools and public libraries. Italians generally express support for the public-school system, and indeed none of the accounts described here makes mention of private schools. Universities are also mentioned, especially in Ostiense which hosts Rome 3 University. Libraries are instead important not only for book lending, but also as a place where university students spend their days studying for their courses. Most of the participants were university graduates, and many mentioned with nostalgia the years they spent in the public library along with other local university-age citizens.

What receives little mention are instead the churches which abound in Rome. Only one citizen made any mention of a parish church, and no church plays a role in the more complex narratives. Post offices are mentioned as a local asset, well-known as locations where citizens pay utility bills and collect their pensions. Their inefficiency is legendary, and complaints directed against post offices are common in local banter. In one case a theater was mentioned as an important public amenity, and there are indeed a number of theaters scattered throughout the urban fringe. Even fountains are mentioned in some narratives as a public utility, although in the periphery they are cited chiefly for their absence. The civic and practical importance of ornamental fountains has an ancient history.

Other features of the urban environment which receive treatment in these narratives are gardens ('giardini') and parks ('parchi'). Green spaces are often wild with few maintained amenities to attract users. Gardens are smaller than parks, usually nestled in densely built districts, especially in the city center. They are generally furnished with playground equipment for children, and benches for use by citizens of all ages. Close access to a maintained garden is an important feature of neighborhood quality, since most citizens live in apartment blocks with limited outdoor spaces. Gardens generally have more intense social activity than parks. Parks are more common in the periphery than the center, although some central parks are quite vast, like Borghese and Pamphili. We will see that parks figure prominently in citizen narratives, and are considered an important asset to the periphery.

For all of its celebration, in no case was the piazza cited as a public amenity (Canniffe, 2012). Most neighborhoods are furnished with some kind of space intended by city planners as the main square, but they often fail to attract citizens for the intended purpose of congregation. The omission of these amenities in our citizen narratives may be an indicator of urban decline, but it could be that they are too obvious to warrant mention. Some neighborhoods, like Ponte Galleria, are so poorly designed as to lack anything resembling a piazza. Another significant omission in the descriptions generated by the workshop are shopping malls. Hanging out at shopping malls is common for younger citizens, while residents falling in the age grade of our group have ambivalent if not negative feelings toward malls. These spaces are held to undermine local retail activities whose presence is so important for community life. A final public amenity which finds its way into these accounts is public transportation, and road connectivity.

An assessment of the strengths and weaknesses of the various neighborhoods is implicit in the narratives, although not easily read since the descriptions often have an ironic or humorous turn. The most consistent weakness is proper waste management, a major concern especially on the western edge of the city with its gigantic landfill. Another concern is atmospheric pollution. The absence of spaces for young people – including university students – is a common complaint. Younger people are drawn to the center of Rome for their evening entertainment, which impoverishes the periphery and festivalizes the center. Security is also a concern, even though Rome has a low crime rate. A final urban element which must be included in this list of semantic features concerns immigrants. Rome has the largest immigrant population of any Italian city. The participants of the workshop were all ethnic Italians, but in one case a story was built around the presence of immigrants.

Intimate tourism

In eliciting ideas about urban life, the seminar encouraged participants to generate an inventory of lexical elements available for the production of stories. The sequence of activities moved from the initial stimulus to think spatially, on to deeper reflection concerning the evocative capacity of the physical environment. The next exercise in the sequence pushed forward the idea of the shared identity between the city and the citizen in an effort to invite participants to rethink the subject-object relationship. Latour talks about the interpenetration between persons and things, showing how meaningless it is to speak of citizens in isolation from the world around them. As we have noted, this new perspective transforms the solipsistic conception of a static individual into 'the moving target of a vast array of entities swarming toward it' (Latour, 2007:46).

The approach to this part of the project was inspired by Luisa Cavarero's observation (Cavarero, 2011) that identity is not expressed by the subject herself, but is instead furnished in accounts provided by the people around her. Cavarero is a philosopher and one of Italy's most prominent feminists. Significantly her approach to feminism has been characterized as dialogical, rather than the monological feminism advocated by some of her colleagues (Re, 2002). Transposed to our setting, the dialogue is between the individual and the surrounding world. Drawing from various sources, Cavarero notes that narrative is a delicate art revealing meaning without committing the error of trying to define it (Cavarero, 2011:10). This is the virtue of narrative studies. Her most important contention is that identity is established not by what one says about oneself, but by third party accounts. An anodyne example is the story of Oedipus: the turning point in his life trajectory is the revelation of identity furnished by a person who tells the story of his unhappy origins. If we wish to understand the real meaning of existence, we need to understand what story can be told about the person involved. I fashioned this idea to the needs of the seminar, predicated on the assumption that places and things can take initiatives and express meaning. Rather than describing the physical circumstances of their urban condition, participants were invited to reveal the way these physical circumstances describe the resident. The exercise aimed to stimulate reflection on agency within the grammar of spatial narrative by creating itineraries of the intimate locations within the neighborhood which identify residents as individuals. To establish a narrative point of view, I asked them to describe the imaginary visitor they were taking on the tour. As with the previous activities, selected results from the exercise were to be included in the theater performance.

The sequential nature of the process, starting with a basic language of the city and building up to increasing narrative complexity, worked well. The intimate tourism exercise allowed citizens to utilize the lexical elements that comprise the city in a more complex narrative syntax. Since these stories illustrate the identity of the speaker it is not surprising that participants fastened on positive features of the neighborhood. In some cases, finding a positive feature was a challenge, a veritable translinguistic activity. The most significant example of this challenge was in Magliana, whose planning shortcomings we have touched upon. Poor planning correlates with a dearth of public spaces beyond streets, minimal sidewalks and building exteriors. Residents claim that ninety percent of the available land area is covered with buildings and the barely adequate road system. Association with the city's most highly publicized criminal gang augments further the stigma of the neighborhood. All these factors are antithetical to any idea of a quality urban environment. Yet there is also something heroic about coming from this area, a badge of distinction justified when citizens are able to overcome obvious adversity through a spirit of struggle and resistance. These are backshadow elements in local narrative about underprivileged neighborhoods.

In the common folk model, the material circumstances of existence are held to impact the quality of personal and social life in a decisive manner. The anticipation is that an area like Magliana with its strongly negative physical attributes - the product of political corruption, speculative growth and what is locally termed 'cementification' – should engender pervasive social deviance. The close nexus between physical conditions and social character can provide an implicit starting point for an intimate tour which is given a poignant turn by introducing a completely unexpected feature of the neighborhood. In this case it is the park, installed in the earthworks in the 1990s (Fig. 5.5). In this narrative the park becomes a new dimension acting as a civic enabler which allows citizens to cultivate the fine art of socialization. The story says much about the importance of parks.

Figure 5.5 Magliana with the raised embankment visible in background. (Photograph taken in April 2020 © Gregory Smith.)

For me the park represents the true identity of the neighborhood; throughout the lives of all the residents it has real value. I remember the afternoons spent as a child playing in the park, skating, and reading the sign posts on the trees so I could learn the species and their characteristics. Public green areas stimulate the imagination of children, create familiarity with nature and thus with life itself. As I grew up, I learned that the park is an area for socialization, where young people can sit and talk, and discover the world.

(Female, Magliana)

Most of the neighborhoods described here are not locally perceived to be underprivileged, and correspondingly the positive idealization of the community is more easily expressed. We see this in Villa Bonelli, a quartiere separated from Magliana by Via della Magliana. This attractive neighborhood is set in the hills which cascade from the west down to the Tiber valley. Most of the buildings are well-conceived middle-class housing units, part of a privileged neighborhood citizens would define as 'civil.' In this setting, the captivating turn of the narrative is that in a civil and privileged neighborhood the highest expression of civility should be found in a critical local amenity which is the public school.

As the first place [establishing identity] I chose my elementary school, the ex Leonardo Sciascia in via Pasquale Baffi. This place is dense with memories, it marks the beginning of my journey in society. This was thanks to the teachers, special and open-minded people even then, whom I will never finish thanking. I also thank my family for choosing this school over others which were closer; perhaps because it was between the Portuense and the Magliana, combining a unique array of experiences and emotions. Still today meeting some of the protagonists of that period, I realize how certain memories are particularly vivid where they were once lived.

(Male, Villa Bonelli)

Schools are presented here as a place to perfect the moral journey through life in the city. In uncivil Magliana we instead discover a silver lining, built ingeniously into the narrative of the neighborhood's failed planning. Both are positively idealized visions of two quite different peripheral realms. The Magliana story in particular echoes the narratives of the 1950s where the intrinsic sacred quality of the individual overcomes adversity and gives rise to a spontaneous community.

As we venture out through the periphery different kinds of negative ideals are occasionally found. This is the case of Ponte Galleria, that poorly designed modern neighborhood comprised of respectable palazzine and villini – essentially small-scale residential buildings – forming an urban environment which the resident described as absurd. Note that the participants were asked to specify whom they were taking on this tour, giving substance to their narrative bias. In this case the author chose a visitor from Germany, no doubt because Germans represent an image of law-abiding rigor, a perfect counterfoil to the chaos of the Italian urban periphery.

Katrina, a young woman of my own age whom I met during a trip to Germany, came to visit me. In travelling from the airport to my home I was overcome with a sense of discomfort, because I realized that I was about to introduce her to a world which she could not even imagine.

(Female, Ponte Galleria)

In these pages I only provide a small number of examples drawn from an exercise developed over several weeks. As a whole these stories reveal the meaning of a physical world where intrinsic civil qualities find expression notwithstanding adversity, and are sometimes even stimulated by its presence. What resonates throughout the accounts is deep attachment to place, and a desire to expose positive meaning which is not obvious when the neighborhood is seen from the outside. The negative mainstream vision of the periphery is a background feature which actors resist by revealing the virtues of the periphery understood especially in opposition to the artificiality of life in the city's privileged areas.

Narrative

The periphery is parataxic in that it lacks a cohesive character imposing meaning on the resident. It is a collection of multifarious elements with little meaning as seen from the outside. But from the inside, this jumble of meaning potential discloses a rich source of opportunity for semantic construction, a kind of assemblage that Rauschenberg turned into an art form in the 1960s. Some writers still describe the periphery as being dull and drab, wrenched by alienation and despair. Such descriptions betray a lack of experience in the periphery. Pasolini was a privileged citizen, but made a genuine effort to immerse himself in the life of the periphery so he could see the city through local eyes, and perceive the periphery's 'desperate vitality.' This is what citizen narratives also allow us to do.

Narrative – what I consider story telling – is different from annals or chronicles in that it almost always contains a plot. Plot consists of the passage from one equilibrium to another, starting with a stable situation which is disturbed by some power or force, in a process which eventually returns to a new equilibrium (Czarniawska, 2012:19). Plot requires the presence of causal forces which can vary infinitely, often revolving around the classic tension between destiny or fate and human ingenuity. In the seminar's final exercise participants were organized in self-selected groups determined by uniformity of residential area. They were invited to write stories set in their local community with any content or structure as long as they were geographically based urban stories. In preparing for this stage of the workshop we reviewed a range of narrative approaches to telling a story. In all cases the value of facts was operationalized by the overall narrative plot. I suggested a simplified approach to emplotment drawing especially from Czarniawska's (2012) discussion of narratology as a research tool in the social sciences. Most narrative structures found in popular media expressed through print, broadcast, or digital media draw from the standard rhetorical techniques employed in western literary tradition. Plot structure selectively highlights variables which are part of shared experience, casting new light on the details that make up the story. These features allow citizen narratives to reveal the range of meaning which can be ascribed to the experience of the city.

Czarniawska (2012) reviews four genres in her study of narrative research. The most common is romance, based on the idea that all creatures and things of the world have a true and deep meaning which is eventually revealed even if faced with every manner of adversity. Tragedy is based on the assumption that humankind is subject to the laws of fate, like Sisyphus with his boulder. No matter how ingenious and committed are human intentions, every effort to achieve their aims is predestined to fail. This genre commonly finds expression in stories about mafia, one of the most successful genres in Italian TV and film. Comedy is when humans are not subject to the laws of fate, but form an organic part of a higher unity, which despite comic complications, works out to resolve everything into harmony. This is the classic format of TV sitcoms of which there are many Italian productions. One of the most popular in Rome is *I Cesaroni*, a sitcom about a family set in the iconic Roman neighborhood of Garbatella (Grignaffini and Bernardelli, 2017). Finally, satire rejects the idea of true meaning in human existence, or rational laws of fate, or some ultimate harmony, using irony to develop a message of skepticism, contradiction and paradox. This is also a common TV and film genre. These four genres are by no means exhaustive of possible narrative styles, and participants were at liberty to use any genre they wished. Among the other narrative styles used in this exercise are modernist ones with no plot structure, or the cliffhanger where things happen but come to no conclusion. The narrative styles could be derived from any source, including literature, film, TV and comics – or other narrative structures utilized in the urban fringe.

A host of narrative techniques is used in constructing the ten final stories, each set in the Roman periphery. Romance is the most common genre, but mafia stories are also present. The noir is common, realized in detective style. Other stories are comedies, romance inflected epics or horror stories. Some of the most charming stories have no plot whatsoever. Various specific themes are adopted in structuring these narratives. Environmental stories are based on struggles to resist the devastation of the territory especially caused by the massive landfill at Malagrotta. Political corruption is a theme, with a suggestion of mafia collusion. Only one story involves a non-European foreigner, treated in an accepting way. More than one story gives prominence to gender, implicitly critiquing the public sphere as a traditional male domain. One story involves a local open-air market, another the local library. Citizen cohesion and the triumph of collective action over the corrupt state provides the backdrop for one story. One of the gendered stories revolves around male intrusiveness in the public lives of women. Another story revolves around the social capital found in an underprivileged community noting the contrast in perception with mainstream culture.

Some literary techniques found in narratives collected in urban environments in the United States are also used in citizen narratives in Rome (Ochs and Capps, 2001). This includes backshadowing, which I have already introduced. Foreshadowing provides clues that help the reader anticipate future developments, and sideshadowing involves the reader as the protagonist thinks their way through a series of events. The plot structure almost always starts with an equilibrium, that is then interrupted by an unexpected event, and finally transformed into a new equilibrium. Umberto Eco (1964) shows how this Aristotelian tradition has found its way into the three-part story so common in Italian popular culture. This is the structure of vernacular story telling in Rome. The first part of the story sets the stage – in Rome called the antefatto. The second part is the dramatic event which gives tension to the story – in Rome called the fatto. The third part is the resolution of the tension created by the fatto – in Rome simply called the fine.

People who know the city will find the stories intriguing. But any reader will appreciate the mastery citizens demonstrate in weaving spatial attributes into a story about the periphery. Here I provide a synopsis of the ten stories and a brief analysis of each. The titles to the stories are those given by the citizens writing them.

Story 1. Guerrilla gardening (Ponte Galleria)

Synopsis. This is an account of guerrilla gardening, a popular engagement in many peripheral parts of the city, where citizens adopt an unutilized green area to establish a community garden. Five young women gardeners are described in the account, with a passion for flowers and horticulture. They are accompanied by ten children on a beautiful sunny day, in an enterprise which also aims to enhance neighborhood sociability. The children are ecstatic about the prospect of being involved in guerrilla gardening. The area is given the fictional name of Rio Galleria. Once the garden is established and the first plants begin to bloom, the children display their name tags before the flowers each is asked to care for. The garden is a place of delight where children play hide and seek, and chase after butterflies that waft through the air.

The sun goes down in the late afternoon, and one of the children is unable to find her playmate, Teo, who had been at her side till then. At this dramatic turn of events the children stand before their respective plants, trembling with fear. The women gardeners spend all night long searching among the vegetation for the missing child. At dawn Teo's plant turns ruby red. A butterfly alights on it and falls dead to the ground, as the roots reach out toward the feet of the surviving children.

This unsettling event recurs daily until all ten children are lost. The only trace of them is their screams, heard every day when the sun goes down, giving way to the moon.

Analysis. This story is in the horror genre. The central characters are women, who often initiate urban gardening. The story establishes a feeling of excitement about being engaged with nature. Guerrilla gardening is no trivial undertaking, it is filled with mystery and surprise.

Story 2. The four shades of the neighborhood (Portuense)

Synopsis. The main character is a retired professor given the fanciful name of Nello Oppana. The time is early morning. The professor is forced by his gout to stay home where he sits before a window that lets out onto a courtyard of what one imagines to be a large residential complex. The story begins at seven in the morning. From his vantage point, the professor is able to observe the entire neighborhood, directing particular attention at the four female neighbors who fuel his sexual fantasies. Each of these neighbors is described in turn. Carrie the redhead walks her dog every morning.

The professor is in a state of sexual excitement as he imagines her nude while caressing her pet. Margaret is a young woman whom the professor has dubbed 'The Virgin.' She greets the professor every day with an innocent gaze, and he imagines her stretched out on a bed, trembling at his every touch. He projects similar fantasies on the other women. He imagines seducing the successful woman lawyer on the hood of her sports car. He fantasizes about possessing a locally resident nurse on a hospital bed, engaging her in perverse sexual practices. Yet, the narrators tell us, each of these women has her own life story, which in no way contemplates the fantasies of the elder voyeur. At midnight, exhausted by an entire day engrossed in these sexual fantasies, the professor retires to bed.

Analysis. The authors of the story are women. There is no plot, only the grotesque fantasy of an elderly peeping tom. The theme is public and private, and the intrusion of the male gaze into the lives of women. Men who dominate public spaces often have a predatory attitude toward women. Public space thus acquires an ambivalent character for women, owing to the violent visual intrusion of men.

Story 3. The landfill and the magic urn (Portuense)

Synopsis. This is written in an epic literary style, replete with fanciful place names, including the location of Rio Galerius which recalls the depressed peripheral area of Ponte Galleria. This is described as a wasteland inhabited by deformed monsters who wander through the dark woods in search of a magical urn. Five women heroes meet here under an ancient oak tree, hoping to gain possession of the magical urn and restore it to the world of the living. The women warriors aim to use the urn in a coordinated effort to bring life back to the valley. But the urn is kept in a grotto protected by a fierce dragon. It is an unequal battle between the women and the horrible monster.

Analysis. The story ends abruptly with the encounter between the forces of good and evil, creating a dramatic suspension of the narrative in a cliffhanger effect. It is a story written by women about their heroic deeds. The underlying theme is environmental devastation, and it is not difficult to guess that the monster represents the corrupt and inept state which has brought material devastation to the western fringe of the city. Suspending the story seems to reference appropriately the ongoing battle between (women) citizens and the Italian state over the protection of the local environment.

Story 4. The controversy of the library (Marconi)

Synopsis. The story starts with an inventory of the services afforded by the neighborhood, from churches to post offices, from libraries to public schools and a host of retail services. The approving mention of ethnic diversity is a clue to the progressive orientation of the narrators. The story is set in 2034, and concerns the public library which is described as having hosted countless generations of Italian university students, and legions of immigrants who mastered Italian using audiovisual materials and courses organized to serve their needs. After disclosing its significant community importance, a dramatic turn is introduced when the reader is told that the library will be closed owing to budget cuts. Indeed, the city authorities have decided to eliminate the library, and give the space over to a development project which involves the construction of elegant high-income housing. A huge crowd of citizens gathers to defend the library, and the police in riot gear arrive to assure its unhindered demolition.

The narrative then shifts to the offices of the local municipal authority where the president is observing the situation via webcasting. Noting the gathering storm, the politician decides to take on the crowd, and summons an armed escort, complete with helicopter surveillance, to accompany him to the construction site. The once-progressive president over the years has abandoned his pony-tail, and acquired a taste for getting his own way. Upon reaching the crowd, he stops and addresses the citizens from the roof of his car. In the harangue, he claims citizens do not understand the value of a project which will provide work for local residents, and promises with time to build a new library. Having calmed the crowd down, the politician returns to his car and uses his smartphone to transmit a text message to a remote location, 'Mission Accomplished.' At the same time a wire transfer is organized from an impressive high-rise office building to an offshore bank account in the Cayman Islands.

Analysis. The theme is political corruption, the topic of countless local stories. The fact that the municipal president no longer has a pony-tail is an obvious allusion to the pony-tail of a prominent local politician, who becomes the brunt of a playful story with a malicious intent. The plot structure is closely aligned to the mafia genre.

Story 5. The Chinese housemate (Marconi)

Synopsis. This is a story told in the first person about a resident who has a new female housemate, one from China by the name of Christ. As soon as the author meets her, he mentions the historic and religious significance of her name. But Christ does not appreciate the comment, and states she is a non-believer. They communicate in English. The home is pervaded by the aroma of Chinese cuisine, in perfect harmony with the rest of the neighborhood which has residents of various ethnic groups. The author describes the process of recycling trash to the Chinese woman, saying that Italy has many problems with waste management owing to the involvement of organized criminality. The author takes her to the local open market where he discovers that Christ does not even know what an artichoke is. Later he takes her to the river park and shows her the bicycle path. At home he explains how global capitalism has created great hardship for ordinary people. The story ends with a stroll in which the author shows Christ a bench which holds special meaning for him, although he is unable to explain why.

Analysis. The group exhibits a clear progressive stance in showing acceptance of the immigrant, and strong connection to the neighborhood with the mention of exact street names in laying out the sequence of events. The choice of a Chinese protagonist allows them to comment on the presence of ethnic diversity which is in fact a characteristic of Marconi. It is a plotless story that establishes a mood rather than creating and resolving dramatic tension.

Story 6. Notoriety (Marconi)

Synopsis. This story is told in the first person about a man who is hired to murder a woman called Valeria; she is a waitress at a bar in Rome's business district (EUR) and is also involved in social activism. His initial plan is to gain admission to her apartment by claiming he is a journalist wishing to interview her about her political activism. Since Valeria is described as obsessed with media attention, the homicide is certain she will fall for this ruse. Before he enacts the plan he follows her through the city in his car, and sees her holding forth in a television interview concerning illegal gypsy camps along the Tiber River embankment. After this he decides that instead of visiting at home, he can murder her in the park. Here he positions himself in such a way as to ask her for help, claiming that he is lock into a gated area contained within the park. He asks for assistance, and when she falls for the trick, he strangles her. The murderer returns home and relaxes, satisfied with a job well done, waiting for the 7 p.m. TV news. The news reports the event, claiming that Valeria must have got into an argument which degenerated into a violent altercation. In the end, the homicide says, referring to the TV coverage, Valeria got what she wanted.

Analysis. The story belongs to the crime genre, told with a hint of misogynism. The style also betrays intolerance regarding social activism, suggesting that violent death is the just outcome for an activist with excessive concern for media attention.

Story 7. The character of a working-class district (Corviale)

Synopsis. The story is set in the kilometer-long public housing project called Corviale. It involves a young woman, Barbara, who comes from an upper middle-class neighborhood. She has a male friend, Dario who lives in the public housing estate, and she is interested in seeing what this stigmatized neighborhood is like. Dario takes her on a tour where she is described as being torn between curiosity and fear. During the tour Dario illustrates the original project which was intended to bring together one kilometer of residents served by a service and retail area which was then never built. Barbara reflects on how different this is from her own neighborhood where so much space is private and little is given over to the public.

Barbara is amazed that so many people can live in such close proximity. While Corviale is surrounded by open green areas, flats are crowded one next to the other, in a cramped condition worsened by poor maintenance. Dario explains that owing to the negligence of the public housing authority, citizens provide whatever maintenance they can by themselves.

On the landing outside the flats runs a series of concrete planters, and from one of these Barbara breaks off a branch of a flowering plant, thinking she will plant the shoot in her own home. Right at that moment a (female) resident steps out of the flat, and asks why Barbara is damaging the plant. Barbara explains that she is being taken on a tour by another local resident to show her what the project is like. Surprisingly, the woman invites the two young visitors into her home. The couple are taken off guard by this display of hospitality, and even more when the smiling woman makes a gift of one of the potted plants she keeps on her balcony.

Barbara takes the plant home, and with time it grows beautifully. Barbara tells all her friends how she came to possess this plant, saying it is better to ask for a simple gift than to acquire an object by stealth.

Analysis. The story is of the romance genre, showing the enduring value of working-class solidarity. The woman visitor comes from a privileged area, and only discovers the importance of shared spaces in this working-class project.

Story 8. The fruit vendor (Magliana)

Synopsis. Another first-person story, this time told by a woman who works as a fruit and vegetable vendor at one of Rome's open markets. She says she likes all the many years spent in the company of her fellow merchants. She especially likes her friend Sabine, with whom she shared dreams every morning for over 20 years. Then one morning, unexpectedly, Sabine fails to show up for work. The author wonders if Sabine had mentioned to anyone her intended absence, saying that Sabine loved her work and was always on the job. She knew each and every client, their family, children, lives and work. Over the years she had learned to read the looks of each client, to know if they were happy or sad. Yet it was impossible to understand if Sabine was happy. Her children were now grown, and she had always dreamed of traveling, starting a new life of adventure. Until then she had only experienced adventure through the stories of her clients.

Sabine failed to show up for work the next day too. The day before Sabine had loaded her van with merchandise, as she did every morning, but then never showed up at work. Sabine's husband gave the alarm, and everyone, clients and colleagues alike, began the search. People phoned the hospital, and traced her route to and from the market. But there was no trace of Sabine.

Days and months passed, and Sabine's husband and children gave up hope of ever finding her. A new colleague took her place at the market, but it was not the same. Years later the postal worker approached the narrator's stand, for what she thought was a purchase. But instead the visit was made to deliver a letter, the first she had ever received at her workplace. Inside was a photograph of a fruit stand, with a beautiful blue sea in the background: it was obviously some exotic location. Unfamiliar fruit trees surrounded the sales stand, along with a multitude of buyers, all with smiling faces. All she could see of the seller were the hands, the hands she recognized to be of her lost friend Sabine.

Analysis. This has the character of a detective story, but ends as a happy romance. The story expresses satisfaction with a public life spent working in a fresh produce market.

Story 9. Violence against a child (Marconi)

Synopsis. This story is also told in the first person. During summer months children are playing in the local park after school. One group of male kids, all about 13, are playing soccer, smoking cigarettes and reading comics. At the same park, but separate from the group, is a boy of the same age with a slight build and small stature. This boy is engrossed in reading a book, as he is most afternoons. The group of children approaches the lone child with a menacing manner as the author witnesses the scene. They throw him to the ground, and break his glasses, while one of the bullies holds out a lit cigarette with the evident intention of inflicting more injury. The narrator who is watching the scene is now prepared for the worst. But contrary to her expectations the victim of the violence snaps up to his feet, and with unexpected agility kicks the threatening hand causing the cigarette to fly into the face of the aggressor.

Analysis. This is also told in a romance style, with an unexpected turn of events and a happy ending which shows that even a timid child with a passion for reading can defend himself in public. The theme is diversity in the park, and the challenges of adolescent life.

Story 10. Pregnant women (Corviale)

Synopsis. This is a story told in the first person by a woman narrator who describes an event involving adolescent girls on their motorbikes. She describes getting up early, having breakfast, and heading off to school – late as always. It is a beautiful warm October day when she finally arrives at school and plants a kiss on the cheek of her best friend. During the recreation period she notes the distress of another friend, Giuliana, who looks terrified and finally breaks into tears saying that she had turned up positive on a pregnancy test. Both girls are 15, and they are encouraged by a third girlfriend to visit a family clinic on the Roman outskirts. The narrator states, in an impetus of womanly solidarity, that they are all pregnant! After the visit to the clinic they climb onto their motorbikes and head home. They get lost on the way, and find themselves unexpectedly at the kilometer-long public housing project (Corviale). They had never seen the building up close before: enormous, gray, sad and dirty. They get off their motorbikes and sit at one of the entrances of the building, on a cement wall. They are as sad as the building itself. Giuliana is particularly distressed, because she has little time to decide what to do. As they are sitting there, fighting back tears, a poorly dressed women with a shopping cart appears before them.

The old woman asks in thick Roman dialect what they have to cry about. Life stinks, she says, but that is all they have.

This encounter causes the girls to realize that they cannot give up hope. They are not even 16, and must learn to deal with mistakes. The author says she learned that what appears to be gray and drab is not what she sees with her eyes, but what she feels in her heart.

Analysis. Corviale is the symbol of human despair, but can still teach a lesson of hope.

High narrative and low narrative

An important issue in the debate on popular culture in the 1960s concerned the relationship between high culture and low culture. The question was whether the popular classes can produce truly autonomous cultural products. Umberto Eco (1964) in a widely read volume explored the relationship between high and low literary creations in the days when Pasolini was still directing his last films on Rome. He offers a disturbing assessment of the quality of popular literature featured especially in comics of those days. He notes that mass readers are essentially lazy, captivated by a simplified narrative format which obliterates historical context in favor of timeless circularity. Such literature denies the relevance of historical experience, and proposes as absolute truth an infinitely reiterated narrative message which is accepted at face value by an entirely uncritical mass audience. This literary format is directed at citizens who live in a highly technological consumer society, lacking all capacity for independent thought or responsible action. This is popular literature in the sense of being consumed by the popular classes, but is hardly a product of their initiative. The consumers of this literature are hetero-directed, lacking any capacity to resist the appeal of mass-produced cultural products.

Pasolini and Eco in those days were separated by acrimonious exchanges. In particular, Umberto Eco accepts the modern conception of language advocated by Ferdinand de Saussure, whereby meaning is governed by a principle of linguistic arbitrariness, with no necessary connection between signifier and signified. Pasolini instead rejects de Saussure, and sees meaning in the language of peasants and citizens of the periphery as emerging from an archaic linguistic source, a timeless reality language. These two positions are irreconcilable. Yet while differing in their views on the nature of language, the two authors converge in agreeing that modern mass cultural products fail to reflect the true culture of the popular classes. The fundamental question for both authors, and for narrative analysis today, concerns the capacity of popular classes for autonomous cultural production. Eco suggests that this capacity is absent. Pasolini instead defends the authenticity of popular cultural products, but denies the presence of any active form of cultural autonomy since these products arise from an unreflecting source rather than from a conscious act of creation.

The narratives generated by the workshop are certainly autonomous creations, understood in relation to mainstream signifying systems which provide little or no support for positive identity in the periphery. In this context citizens can either accept a negative reading of their condition, or find new ways of giving meaning to their particular status. These narratives are not hetero-directed, nor are they expressions of timeless reality language. They are autonomous cultural products which undoubtedly reflect Pasolini's redefinition of the periphery, turning membership into a mark of distinction rather than a demerit. Yet the issue is complex. Fortunately, there are alternatives to the antinomy between hetero-direction and pure autonomy. The best alternative may be that offered by Michel de Certeau (1990), who believes that while citizens can be dominated by the products of the mainstream system, they then use these products in an autonomous way. They are dominated by commercial value production, but at the same time appropriate this value for new uses (Certeau, 1990:xliii). Thus, we might consider three options in assessing the Gramscian question of the autonomy exhibited by popular classes in the area of cultural production. Most likely these three different forces are combined differently in different settings. This is the topic to which we turn in the next chapter.

Notes

1 This is especially linked to my work with the Cornell University Rome Workshop where I directed for years undergraduates in explorations of Roman peripheral neighborhoods. Cognitive mapping is a key feature of these investigations (Smith et al., 2014).

2 The creation of municipi is part of a long process of administrative decentralization which began in the 1960s. By 1990 there were 20 municipi, then in 1992 one broke off and became an independent township. There were only 19 municipi left, numbered from one through 20, skipping 17. The law changed again in 2008, when the number of municipi was reduced from 19 to 15. The functions of municipi are set out in City of Rome's ordinance 122 of 2000.

3 City ordinance promulgated by the special commissioner of Rome, number 2453 of September 13, 1961.

4 City ordinance 2983 from July 29/30, 1977.

5 Here we must recall Bellicini's study of the 1990s when he proposed that some 200 micro-quarters should replace the many existing territorial units (Bellicini, 2001). Yet the city continues to change, and his proposal for a new system of spatial classification was never adopted.

Bibliography

Augé, Marc (1995) *Non-places: introduction to an anthropology of supermodernity.* Translated by John Howe. London: Verso.

Bal, Mieke (2003) 'Visual essentialism and the object of visual culture,' *Journal of Visual Culture*, 2:1:5–32.

Bellicini, Lorenzo (2001) 'Le "microcittà"' e il nuovo piano regolatore,' *Urbanistica*, 116:198–199.

Black, Rachel E. (2012) *Porta Palazzo: the anthropology of an Italian market.* Philadelphia: University of Pennsylvania Press.

Canniffe, Eamonn (2012) *The politics of the piazza: the history and meaning of the Italian square.* London: Ashgate.

Cavarero, Luisa (2011) *Tu che mi guardi, tu che mi racconti. Filosofia della narrazione.* Milan: Feltrinelli.

Cellamare, Carlo (2014) 'Ways of living in the market city: Bufalotta and the Porta di Roma shopping Centre,' in Marinaro, Isabella Clough and Bjørn Thomassen (eds.) *Global Rome: changing faces of the eternal city*, pp. 143–155. Bloomington: Indiana University Press.

Certeau, Michel de (1990) *L'Invention du quotidien, i. Arts de faire.* Paris: Gallimard.

Clifford, James and George E. Marcus (eds.) (1986) *Writing culture: the poetics and politics of ethnography: a School of American Research advanced seminar.* Berkeley: University of California Press.

Czarniawska, Barbara (2012) *Narratives in social science research.* London: Sage Publications.

Denzin, Norman K. (2010) *The qualitative manifesto. A call to arms.* Walnut Creek, CA: Left Coast Press.

Denzin, Norman K. (2018) *Performance autoethnography. Critical pedagogy and the politics of culture.* London and New York: Routledge.

Eco, Umberto (1964) *Apocalittici e integrati.* Milan: Bompiani.

Faulkner, Sandra L. (2009) *Poetry as method. Reporting research through verse.* Walnut Creek, CA: Left Coast Press.

Grignaffini, Giorgio and Andrea Bernardelli (2017) *Che cos'è una serie televisiva?* Rome: Carocci Editore.

Ingold, Tim (2000) *The perception of the environment. Essays in livelihood, dwelling and skill.* London and New York: Routledge.

Ingold, Tim (2007) *Lines: a brief history.* London and New York: Routledge.

Jacobs, Jane (2016) *The death and life of great American cities.* New York: Vintage.

Landinin, D. Jean and F. Michael Connelly (2000) *Narrative inquiry. Experience and story in qualitative research.* San Francisco: Jossey-Bass.

Latour, Bruno (2007) *Reassembling the social: an introduction to actor-network-theory.* Oxford: Oxford University Press.

Lefebvre, Henri (2006) *Writings on cities.* Edited and translated by Eleonore Kofman and Elizabeth Lebas. Oxford: Blackwell.

Lynch, Kevin (1960) *The image of the city.* Cambridge, MA: MIT Press.

Lynch, Kevin (2002) *City sense and city design. Writings and projects of Kevin Lynch.* Edited by Tribid Banerjee and Michael Southworth. Cambridge, MA: MIT Press.

Marcus, George (1997) 'The uses of complicity in the changing mise-en-scène of anthropological fieldwork,' *Representations*, 59:85–108.

McFarlane, Colin (2011) *Learning the city. Knowledge and translocal assemblage.* Sussex: Wiley-Blackwell.

Modan, Gabriella Gahlia (2007) *Turf wars. Discourse, diversity, and the politics of place.* Oxford: Blackwell Publishing.

Ochs, Elinor and Lisa Capps (2001) *Living narrative. Creating lives in everyday storytelling.* Cambridge, MA: Harvard University Press.

Piccinato, Luigi (1993) *Urbanistica medievale.* Bari: Edizioni Dedalo.

Re, Lucia (2002) 'Diotima's dilemmas. Authorship, authority, authoritarianism,' in Parati, Graziella and Rebecca West (eds) *Italian feminist theory and practice. Equality and sexual difference*, pp. 50–74. Madison: Fairleigh Dickinson University Press.

Rossi, Aldo (1984) *The architecture of the city.* Cambridge, MA: MIT Press.

Ruggeri, Deni, Chester Harvey and Peter Bosselmann (2018) 'Perceiving the livable city: cross-cultural lessons on virtual and field experiences of urban environments,' *Journal of the American Planning Association*, 84:3–4:250–262.

Scollon, Ron and Suzie Wong Scollon (2003) *Discourses in place: language in the material world.* London and New York: Routledge.

Sitte, Camille (1979) *The art of building cities: city building according to its artistic fundamentals.* Cambridge, UK: Ravenio Books.

Smith, Gregory (2013) 'Narrative in place: perspectives on Pasolini's Rome,' in Smith, Gregory and Jan Gadeyne (eds.) *Public space in Rome through the ages*, pp. 277–299. London: Ashgate.

Trancik, Roger (1986) *Finding lost space: theories of urban design.* Hoboken: John Wiley and Sons.

6 Urban arts and the sacred

Cultural autonomy

Sophisticated observers note the importance of rhythms and sounds in urban life. Kevin Lynch uses the experience of passing over the San Francisco Bay Bridge to illustrate the point, a passage which hints at the organization of melody in classical music with its introduction-development-climax-conclusion sequence (Lynch, 1960:99). Henri Lefebvre (2019) theorizes a discipline he calls rhythmanalysis, embracing the qualitative and quantitative aspects of human experience, bringing within a single purview a wide range of interconnected phenomena, including music. Music can provide a metaphor for understanding spatial organization, enhancing our experience of the city. And knowledge of the city may in turn enhance our understanding of certain types of music. In these settings music is an autonomous cultural practice constrained by the physical circumstances of its creation.

Poetry is also an art form with strong attention to sounds which can enhance the portrayal of spatial experience. We see this in Giacomo Leopardi's poetic descriptions of village life in 'Il sabato del villaggio' (Saturday in the village), with its reference to the sounds of the blacksmith's anvil. Pasolini transposes this rendering of sonorous experience to the description of the Roman ghetto in the 1950s with its voices, a blasting radio, a passing automobile. From a cosmological standpoint both of these poets reference the fact that the creation of the universe originates from sound (Banda, 1990:179). Cross reference between sound and spatial experience is still important in the arts today. Moneyless is one of the street artists involved in the 2013 Ostiense underpass project. He has an art piece in the housing project at Tor Marancia which residents claim he realized while listening on earphones to music with a heavy bass and drum rhythm. The mural captures these sounds in visual representation (Fig. 6.1).

Music today plays an important role in the urban arts, a term I use to reference street art and certain types of music that celebrate urban life, particularly but not exclusively rap.[1] Italian rap has identical historical roots to its street art, having both originated in the social centers which sprang up all over the country in the 1980s (Mudu, 2014). These centers represent an

Figure 6.1 Moneyless's mural at Tor Marancia. (Photograph taken in April 2020 ©
      Gregory Smith.)

alternative to Italian mainstream culture, and while most are concerned to
promote cultural and political activities, some are residential and currently
house 5,000 families in Rome (Grazioli, 2017). Squatters see themselves as
the vanguard of a movement against the rent economy. Land owners, builders and banks, they claim, cause the uncontrolled growth of the city, with
their decaying peripheries, poor and inefficient services, shortage of public
housing, evictions for failure to pay rent or the inability to make mortgage
payments (Lamanna, 2019). Social centers resist the logic of the neoliberal
city, and are in some ways an expression of the right to the city (Montagna,
2006). One of the oldest social centers in Rome is Forte Prenestino, founded
in 1986. Here one finds early expressions of graffiti and Roman rap (CSOA
Forte Prenestino, 2016). The Roman street artist who goes by the name of
Hogre was active at Forte Prenestino already in the 1990s. He claims that
Italian street art has its roots in the 1960s, especially in the work of the
neosituationists. This was a Marxist-inspired group allied to Guy Debord,
which resisted the transformations entailed by the neoliberal political and
economic system (Tommasini, 2019).

Contemporary art forms are global, often reterritorialized to give greater impact to their message by highlighting local attachments (Hennessy, 2015). It is a modern deixis of resistance which finds expression in both visual and musical art forms. Pasolini makes expert use of music to capture the spirit of resistance in his poems and films. He features graffiti in the same media. In 'The tears of the excavator' he describes the words 'Viva Mexico' written on the decrepit surface of a Roman monument, perhaps to signify some obscure translinguistic operation (Pasolini, 1976:103). A visual image of graffiti is seen in borgata Gordiani in the film *Accattone*, where the audience reads scrawled on the wall, 'Voglio una casa decente' (I want a decent home). This shows the extent to which the city was a site of contestation already in the 1950s, a canvas on which underprivileged citizens could leave their mark.

These urban arts are typically embedded in geographically discrete communities, and consciously promote a form of autonomy which resists the deracination of modern global culture. Historically in Italy the question of cultural autonomy has fallen within the purview of 'demological' studies exploring the interface between subaltern and hegemonic classes. After World War II these studies gradually experienced decline having failed to take into account the problems of globalization, digital communication and mass consumption (Dei, 2013:84). Pasolini also wrote extensively about subaltern cultural autonomy, taking Gramsci as his starting point. Gramsci notes that many elements of popular culture have a bourgeois origin, and suggests that they are imposed on subaltern classes. Pasolini instead expresses the view that even when elements are borrowed from a hegemonic source, in the new popular setting they become part of a fixed and undifferentiated traditional culture (Pasolini, 1955:XLI).

Gramsci has a negative view of folk culture, considered an incohesive collection of fragments derived from higher classes (Dei, 2013:89). According to this interpretation, the subaltern classes are unable to formulate an organic expression of their worldview because they lack the community cohesion required to do so. The only area where they are able to fuse together an organic cultural system is in the dialects Gramsci believed should be preserved through public school programs. Pasolini, instead, cultivates a romantic view of popular culture especially viewed through its poetry. Poetry captures the archaic spirit of the Italian people, which he sees as being sharply different in the north of the country as compared to the south. In the Italian north, poetry and song are narrative, while in the south they are lyrical (Pasolini, 1955:XXI). In the north words are chosen in a conscious manner, while in the south accidental words give rise to a poetic concept. In the south the popular author is an elegant artist rather than a poet, superior in this art form to any other people, the Greeks excepted. The poetry of the south fashions rather than creates, expressing a culture which surfaces spontaneously from archaic roots.

According to Pasolini, if there is autonomy in local cultural production it is driven by unconscious membership of an archaic community rather than by the conscious action of a reflecting actor. This is the position expressed in the description he provides of dialect poets in an essay written in the 1950s: '[The dialect poet's] greatest ambition is to dissolve into anonymity, become the unconscious demiurge of the genius of the people of his city or town, spokesperson for absolute local happiness' (Pasolini, 2009:61). The idea of anonymity and the attendant void of individual autonomy is also expressed in the notion of 'speaking subjects.' He uses this concept in the description of his Friulian research first published in 1947 and later included as an appendix to the 1979 Einaudi version of *Ragazzi di vita*. It illustrates the thinking behind his ethnographic method. Here too he speaks of the anonymity of the speaker, and linguistic mimesis, presenting his idea of dialect as representing a language of reality directly connected to humankind's archaic roots. This conviction is the foundation for his polemic against de Saussure's idea of the arbitrariness of the linguistic sign. In Friuli and the Roman periphery, Pasolini documents the reality of popular culture by immersing himself in the language of that reality. Popular poetry, like popular language, lacks self-awareness, existing despite the speaker's intentions as a direct expression of archaic truth and beauty (Dei, 2013:97).

Put in these terms, the poetic creations of southern peasants and the Roman underclass are spontaneous and unselfconscious, expressing the passive relationship these citizens have with their own cultural products. Precisely since their generation is unreflecting, these creations, and the entire cultural system which supports them, are susceptible to distortion by the forces of modernity. The only action able to modify the system without destroying its foundation is the power of formal poetics through the force of translinguistics (Pasolini, 2005:198). This is the antidote to the system of commodity production, rupturing the Pavlovian connection it creates between signifier and signified. Formal poetic action transforms our vision in a positive way, introducing an element of transparency to the meaning system which is absent in archaic reality language. Commercial signs instead distort meaning for artificial aims. Translinguistics opposes this obfuscation of meaning, rescuing in the process the popular classes from cultural decimation. To be sure, this formal poetry is not a spontaneous or accidental cultural product, but the work of an inspired poet of which Pasolini could claim to be a distinguished example. He could change the vision of the world in a positive way, helping preserve a culture whose carriers had little capacity to control. The best example of translinguistics is certainly the modification of the periphery's image from a sign of discredit to a sign of the sacred. This transformation provides the foundation for resistance to the forces of control, and the autonomy of subaltern cultural production.

It must be noted that a literal reading of Pasolini's linguistics can be construed to betray a distinctive elitist element. Asor Rosa in the 1960s discerned this bias, and wrote Pasolini off as a reactionary. Even contemporary specialists of Italian popular culture see Pasolini's characterization as the 'most radically antipopulist statement in Italian literature' (Dei, 2013:95). Yet notwithstanding this elitist element, the image Pasolini cultivated as a champion of the poor is what persists today in the periphery. What reaches the general public are not his theoretical writings, but the detailed images and sympathetic treatment he provides of marginal classes. The rich documentation found in his Roman films and writings substantiates his claim to be an advocate of the excluded. Within this wealth of information, we also find examples of the use of poetry among the popular classes, such as in the film *Mamma Roma* where a prisoner recites Dante's *Inferno*. He is an unschooled Dantist, and his declamation is replete with inaccuracies. The episode demonstrates that the popular classes borrow from high culture, and then redeploy the borrowed elements in a novel way reflecting the new conditions in which that cultural element is inserted. This usage distorts the original poetic work while revealing the universality of its message, showing how the popular classes appropriate high cultural elements in a way which belies the literal reading of Pasolini's theoretical reflections.

Song receives more treatment than poetry in Pasolini's novels and films, where we see many instances of citizens bursting unexpectedly into song. Rarely do we know what song is involved, but we are allowed to understand that it is an expression of commercial music filtered down to the underclass. In *A violent life* the protagonist and his friends venture out to a girlfriend's house in Garbatella, and from beneath her closed window offer a serenade of songs accompanied by guitar. The tunes performed include American classics, like 'Oh Suzanna' and the Platter's 1955 hit 'Only you.' He renders the local interpretation of these songs by transcribing the Roman pronunciation of key words, such as *Oli U'*. This is an instance of the diffusion of commercial culture among the popular classes, the melodic equivalent of the 'American shirts' that abound within the same community. These acquired elements are assimilated into the general character of the new social setting. The commercial traits incorporated into popular culture in the 1950s are not only expressions of 'Americanization.' *Mamma Roma* features importantly an Italian hit from the 1930s, 'Violino Tzigano,' showing how the popular classes adopt traits from Italy's own commercial song culture. As de Certeau notes, the popular classes may be dominated by these cultural artifacts, but then appropriate them for their own cultural purposes (Certeau, 1990).

Pasolini's films also document what can be considered true folk features of Italian culture. This is found for instance in the film *The Decameron* which draws heavily from recordings made by Alan Lomax in the Italian south in the 1950s. It is perplexing why the film should not credit Lomax as the source for these recordings, the use of which Lomax discovered by chance during a screening in New York (Plastino, 2008:63). Perhaps Pasolini believed these recordings were a witness to a reality language which belongs to humanity rather than to the ethnomusicologist who collected them.

From the standpoint of folk music, we have detailed treatment in *Mamma Roma* of Rome's most famous example of popular song, the stornello. Stornelli appear in the opening scenes of the film, when the protagonist engages the diners at a wedding feast in a competition using this peculiar form of dialect poetry. Pasolini had written about stornelli in his early study of popular poetry, claiming they were one of the oldest song forms in Italy (Pasolini, 1955:XXXV). The formal characteristics of the stornello are well established, consisting of any number of stanzas, each with a predetermined structure (Savona and Straniero, 1994). The first line is comprised of five syllables, with an accent on the fourth, and usually ending with the invocation of a flower. This line is followed by two others, each composed of hendecasyllables. These three lines comprise the stanza, following an invariant meter whose content is furnished at will by the singer. The singer usually draws the topic for the stornello from some locally evident circumstance. In a congregation like the setting of a meal, each singer stands up in turn to deliver the improvised lines. These exchanges often have a strong element of 'sfottò,' the chafing banter common among citizens in settings where they are considered peers. The dinner in this case celebrates the wedding of Mamma Roma's pimp (Carmine) to another woman. Mamma Roma sees this event as a liberation.

> **Mamma Roma**
> Fior di gaggia
> Quando canto io canto con allegria,
> mo' se io dico tutto rovino 'sta compagniaaa!
> **Carmine**
> Fiore de sabbia
> Tu ridi, scherzi, fai la santa donna,
> e invece in petto schiatti da la rabbia
> [**Mamma Roma**
> Flower of fools
> When I sing I sing with joy,
> But if I say everything, I'll ruin the party!
> **Carmine**
> Flower of the sands
> You laugh and joke, with the air of a holy woman,
> But inside your heart explodes with anger]

(Pasolini, 2006:243)

Stornelli were a common feature of working-class gatherings in the 1960s and 1970s, and are today rather rare. Pasolini claims that stornelli fashion together accidental words in an elegant art form, the direct expression of the archaic sources from which this poetry originates. But whatever the source of their creative force, stornelli certainly represent an autonomous cultural form.

The reading of Pasolini as an elitist seems to be lost outside of specialist circles, and what prevails is the memory of a champion of ordinary people. This is the case of a contemporary singer songwriter who claims to be the heir to Pasolini. Detailed analysis of this claim, and the corpus of work that supports it, reveals a current reading of Pasolini and illustrates his continued relevance today. The singer songwriter is Vinicio Capossela. This analysis lays bare what can be characterized as a postmodern reading of Pasolini. This is unusual since Pasolini is almost always construed to be modern but never postmodern. A critical element which keeps Pasolini in the modernist camp are his fixed spatial reference points. This reading is consistent with his geographical essentialization of the periphery.

Vinicio Capossela: The new Pier Paolo Pasolini of Italian song and literature

Aldo Luigi Mancusi (2012) writes that the radical blog *Brutalcrush* is a place for readers of Bukowski, admirers of musicians like Nick Cave and Tom Waits. It is also a venue for discussing Vinicio Capossela. Indeed, in this site Mancusi characterizes Capossela as the 'spaghetti and mandolin' version of Tom Waits. The early Capossela waxes lyrical about wine, beer, women and bars. According to Mancusi, the later Capossela was ruined by Rome's progressive intellectual establishment, especially after the release of his CD *Ovunque proteggi* (*Protection everywhere*) (Capossela, 2006). The release was celebrated at Rome's *La Casa del Jazz* (*House of Jazz*), itself a product of Rome's mainstream progressive forces; this, according to Mancusi, was the negative watershed in Capossela's artistic career, where posturing leftist intellectuals dubbed Capossela the 'new Pasolini.' Reference by Mancusi to Pasolini does not denigrate the great poet, but rather the modern uses of the memory of an intellectual and artist who has become the political capital of the Democratic Party. Vinicio Capossela himself claims to draw inspiration from Pier Paolo Pasolini, while distancing himself from Italian mainstream politics. Pasolini's eclectic heritage allows for many interpretations which separate as well as unite his readers. Capossela continues even in recent years to draw from Pasolini's heritage, and in 2019 released a video clip called 'The poor Christ' which features the same actor playing Christ in the film *The Gospel according to Matthew* (Pasolini, 1964). And just as Pasolini's film received flattering reviews from the Catholic Church, so the video clip 'The poor Christ' receives approving mention in an important Catholic publication.[2]

A major watershed between Pasolini and Capossela is the divide between modernity and postmodernity. Pasolini's literary modernism is made up of juxtaposition, seepage and misappropriation, articulated within a fixed territorial framework. This is the stuff of a modernist assemblage. With Pasolini we are in the 1960s, when the city had at least the potential to resolve the contradictions of modernity, to reterritorialize deterritorialized process.

Postmodernity further fragments the world, adding fresh challenges to the global city. Capossela is part of this new universe.

Pasolini was born in Bologna in 1922, and later became the bard of Friuli and of the Roman periphery. From a linguistic standpoint, he is unusual for his generation, having grown up using the national language rather than a local dialect (De Mauro, 1999:79). This circumstance may have engendered his deep passion for local linguistic forms. Capossela was born in Germany in 1965 of parents who had emigrated from southern Italy. We can imagine that his first knowledge of Italian was based on a dialect. Yet his songs and writings make use of standard Italian. He later returned to Italy and settled with his family not far from Bologna. It was here that the songwriter established his roots as a performing artist, and it is from here that he set out into the world. Capossela is mostly known as a songwriter, but is also the author of volumes which express sensitivity to the spatial features which weave their way through Pasolini's writings. There are many similarities between the two artists, yet many differences as well.

Capossela shares with Bologna's great poet a fascination for Greece, although explored as much for its contemporary condition as for being the source of western civilization. Among his books we find the volume entitled *Tefteri* (Capossela, 2013), whose focus on the resistance the Greek people exercise against the global finance system reveals Pasolini's inspiration. The book discusses a musical form known as *rebetiko*, taken from the Turkish word *rebet* meaning rebel (Capossela, 2013:38). This music bears witness to the continued revolutionary spirit of the Greek people, although modern Greece is losing its traditions which include a strong spirit of opposition. He advances this claim by making reference to the metaphor of Greece having lost its smashed plates, a transposition he draws explicitly from Pasolini's claim that Italy had lost its fireflies, namely its innocence as a peasant society (Capossela, 2013:13).

Tefteri explores the contrast between two popular Greek musical forms, one the *demotikì,* the other the *rebetiko. Demotikì* is a national musical form, while *Rebetiko* is the music of the one and a half million refugees generated by the Greco-Turkish war. Capossela claims that *rebetiko* stands in metonymic representation of all refugees generated by modern global capitalism. In Greece the success of the musical form was such that in the 1920s a single *rebetiko* album sold more copies than the number of record players then in existence. *Tefteri* describes the celebration of drugs in *rebetiko* song, especially the hashish once consumed in the eastern territories. *Rebetiko* celebrates Greece's classical roots, and in one song Caronte, the god of Hades, carries hashish to the dead souls. Capossela claims that this is an apt symbol for geographical displacement, for people confined beyond a physical barrier, a metaphor for crisis in a world characterized by spatial flux, separated by artificial barricades.

Capossela makes restlessness the hallmark feature of his artistic methodology (Salis, 2013). Restlessness forces him to move from one location to another, and among various artistic media, drawing from a rich range of sources to create innovative fusions of sounds, concepts and places. *Tefteri* is a travel journal full of personal assessments, and descriptions of living personalities. The book shares with Pasolini the desire to engage as an artist in social and political commentary, to unite past and present, poetry and minute description. Like Pasolini, he believes the past and the present are fused together in a continual process of eternal return, mentioning in this context the idea of hierophany (Capossela, 2013:145). He also puts the idea of the sacred at the center of his artistic production. In an interview released during a tour of Sardinia, Capossela illustrates how the island's primitive territory draws together nature and archaic human life.[3] Here he speaks of the sacred, specifying that he is not a believer yet is inspired by Scripture. For both writers a horrendous future is in store, with the loss of the sacred which is no longer experienced as a loss (Conti Calabrese, 1994:9). Capossela's performances celebrate epic stories about the sacred and the beginning of the world. The sacrality of place is seen when Capossela claims that the performance of *rebetiko* in its native setting of the tavern is the musical equivalent of the Eucharist. This is music for rebels who are sacred precisely because they live beyond the reach of the dominant system of social representation.

The liner notes to *Ovunque proteggi* (Capossela, 2006) acknowledge a debt to Pasolini. The songs in this CD make abundant reference to flesh and blood, reminding us that our spatial-temporal presence is expressed through our physical body. This is a form of reterritorialization exercised on the most intimate scale. He draws special inspiration from the Easter celebration, in particular the popular Easter enactments celebrated in Sicily when Christ dies as God and is resurrected as man. In its popular expression this ritual reveals man's joy at having survived the transition to divinity, underscoring the fact that we are all sacred. The individual is sacred, but so is the world he inhabits. Capossela claims to be an animist, and believes that space has a spirit which he feels during his performances (Salis, 2013). This is his idea of hierophany, where ritual binds citizens together, joining social, physical and spiritual dimensions.

Live performances for Vinicio fuse together the sacrality of language with that of place. Conti Calabrese shows that for Pasolini too humankind in its natural state lives the sacred truth of language in the physical setting of everyday ritual (Conti Calabrese, 1994:85). A key figure in Pasolini's film *Medea* is the Centaur who states that everything is sacred, and himself blends together humankind's archaic mysticism with utilitarian rationality. Failure to resolve the two is the foundation for Greek myth. In Pasolini's thought the sacred and the profane must be harmonized: archaic civilization must not be 'forgotten, despised and betrayed' (quoted in Shapiro, 2013:113). Capossela too tries to preserve archaic mysticism, and in the early 2000s spent several years experimenting the relationship between the spoken word and the mythical force of different spatial settings. This experimentation led to the publication of a book, *Non si muore tutte le mattine* (*You don't die every morning*) (Capossela, 2004) which in turn became the foundation for a series of theater readings with musical accompaniment. The experimentation led to the CD *Ovunque proteggi* (Capossela, 2006), containing songs conceived and recorded in diverse parts of the Italian territory. It is a labyrinth of places, capturing the profound meaning arising from the interaction between the immutable sacred and the transience of mortal flesh.

Capossela falls into a progressive songwriting tradition that in Italy has its roots in the 1960s as a response to commercial pop music. Some groups already in the 1950s represented this critical approach, but the real heyday of social commentary through song belongs to the 1960s, especially with singer songwriters like Francesco Guccini and Fabrizio de André (Borgna, 1992). Both of them encouraged Capossela in the early stages of his career. Guccini is most clearly identified as a communist singer songwriter, although radical critics accuse him of 'selling out' in the later stages of his career (Mancusi, 2012). A new season fusing radical politics and song was inaugurated in the 1990s, with Italian groups which blended together American and British rap, combined under Jamaican inspiration, to invent Italian rap (Mitchell, 1996). Radical content, with a clear Marxist inflection, is a distinctive characteristic of the genre, along with the use of dialect instead of Italian language to capture specific features of local cultural geography. This entirely new sonority in its origin is closely associated with the southern Italian group *Sud Sound System*. They were the chief innovators of an Italian musical genre called Tarantamuffin, a fusion between the Italian tarantella and Jamaican Raggamuffin. Tarantella expresses the spirit of the south, especially Basilicata and Puglia. This dramatic collective ritual had such impact on Pasolini that in homage to the people of these regions he filmed *The Gospel according to Matthew* in southern Italy rather than in the Holy Land. Tarantamuffin reterritorializes on sacred soil an international art form.

While Capossela celebrates the sacred value of the south, for him the spatial definition of marginality is no longer clear. Rather than seek a fixed location to experience alterity, he seeks a sacred landscape through life on the road, slumming around in the style of the American beat generation. Capossela is not a *cantastorie*, the storytelling songwriter, but instead is a *viaggiastorie*, a traveling storyteller (Giardinazzo, 2009). He is a traveler who picks up hitchhikers and incorporates their stories into his songs, reveling in the bowels of a world which functions as a universal motel. Travel allows the discovery of a distant territory in one's own land lived as a remote location. In this form of metempsychosis the world is borderless, and travel is the high road to wisdom (Giardinazzo, 2009:114). Capossela's travel philosophy borrows from Pasolini's spatial sensitivity. Yet he challenges the idea that modern Italian has entirely obliterated the infinite linguistic variety rooted in a preindustrial past (Giardinazzo, 2009:169). Diversity persists especially in the south notwithstanding colonialist and racist stereotypes which deny it a dignity worthy of its past.

Capossela has a fascination for the Italian south, a land fragmented and united by the Mediterranean (Giardinazzo, 2009:156). This is Italy's other, the south of de Martino (1961), so different from the north that it represents a transoceanic journey within a single peninsula. It is an arcadia of historic suffering, of mysticism and magic, the land of the tarantella with its ancient roots and modern vitality. Those afflicted by the tarantella cannot stay put in one place, imposing migratory action which has characterized Capossela's entire artistic production. In Capossela tarantella not only establishes the rhythm for music, but also constitutes the inspiration for a fusion between life and art. His fascination for tarantella first appeared in the 1996 CD *San Vito*, a turning point in his career when his concern for movement became firmly established (Piro, 2010:45). Fascination for movement continues in *Canzoni a manovella* (*Songs with a crank*) (Capossela, 2000), a CD dedicated to the automobile. The car is the instrument of an artistic process and a lifestyle, as captured in the documentary film *Live in Volvo* (Capossela, 1998).

Giardinazzo (2009) explores two parallel movements within Italian literature and music. One originates with the Italian progressive rock group called Area, the other with Pier Paolo Pasolini. They represent in their respective ways musical trends leading up to Capossela. Both currents express anti-conformist responses of the late 1960s and early 1970s when consumer culture took irremediable hold of Italian society. In this transition it became impossible to speak of 'popular culture' in the traditional 19th-century fashion. The individual was no longer a consumer, but was consumed by the system. Even the popular songs that emerged in the 19th-century process of nation-building – the Risorgimento – were produced by educated classes who celebrated their own idea of Italian popular culture (Giardinazzo, 2009:105–106). In this period the romanze and arie of the south began to acquire commercial value, most famously the romanza 'O Sole Mio.' According to this view, the dominant classes continued to create and disseminate popular music on up through the 1960s, with an ever-stronger commercial element. The process culminated in the 1960s with a song form known disparagingly as the canzonetta (little song). These little songs were adopted by the bourgeoisie, and through dissemination by radio and cinema filtered down to the popular classes. The deathblow to anything resembling popular culture came with the advent of television in the mid-1950s, destroying the peculiarity of the preindustrial world that had so fascinated Pasolini (Giardinazzo, 2009:116). While in Pasolini's day some forms of true popular song persisted, especially the stornello, Pasolini believed the only way to release song from commercial control was by involving inspired poets like himself in producing songs. But the clannish character of the music industry made this impossible, and in his career Pasolini wrote only one tune. It is about a prostitute in Testaccio and entitled 'Valzer della toppa' (Drunken waltz) (Borgna, 1992:271). It tells the story of a prostitute who drinks too much and dreams she can reacquire her purity by becoming a virgin once more. Linkage to the broader theme of spiritual corruption gives the song a translinguistic quality alien to any commercial song.

Pasolini claims that the sacred unity between words and sounds has been lost in modern commercial songs which are 'foolish and degenerate' (quoted in Giardinazzo, 2009:109). Capossela echoes this sentiment when he mentions his love of public readings, even in languages he is unable to comprehend fully, precisely to savor the sonorous quality of spoken words (Piro, 2010:27). The sound of spoken language captures humankind's archaic condition. Like Pasolini, Capossela cultivates a passion for humankind, not in the abstract, but in the corporeal expression of daily life. His travels are full of tangible objects which repel the void of abstract experience. Words are linked to places in an intimate world expressed through a sonority refined over years of experimenting with live audiences. These performances constitute poetry en plein air. Capossela too seeks a precedent in Giacomo Leopardi, and in a modern interpretation of the shepherd who wanders through Asia, he drifts through the modern world in a Volvo. As in the Dadaist walking strategy, life is lived in motion because this is the only way some things can be understood (Carreri, 2006). Vinicio captures the specific features of Italy's landscapes. Some songs are inspired by the irregular landscape of the Italian badlands (calanchi), while others arise from a landscape strewn with the twisted scotch broom (ginestra) of the Italian south.

Postmodernism is a watershed in western history, the end of any claim to objective truth (Lyotard, 1979). Place in these conditions becomes a kaleidoscope of shifting meaning, utterly opposed to the fixed reference points of Pasolini's geography. Postmodernity fragments the world, creating drifting multilayered hybridity. Pasolini's parataxis and juxtaposition are part of a stable physical environment. Space for Capossela is instead a labyrinth which can be travelled in any direction, a territory with no shared meaning and no fixed confines, a fragmented world, a terrain of competing forces lacking any form of consensus. The very nature of place varies according to the multiple practices which engage diverse citizens in a single location. Spatial praxis shifts endlessly. Quantity expresses the infinite meaning potential of place, which can only be domesticated and rendered comprehensible through the actions of a spatially referenced individual. This is the sense behind Capossela's constant travels, his story collecting, his live readings. Verbalizing experience in place is the only way to compensate for the compression of time and space.

For Capossela objectivity and rationality are no longer the fixed cardinal points of the world. The grand narratives are dead; to understand the world one must first deconstruct existing meaning systems to open the way for a radically new understanding. Postmodern rationality is the rationality of a particular combination of words and experiences expressed in a given place at a given moment in time. Postmodern signifiers accept that multiple meanings can coexist within a single temporal and spatial axis. This is different from the modernist approach which describes the world in a vision which can inspire a rational response to its wayward drift. Capossela lives in a more complex universe than that found in *La ricotta*, where narrative confirms the physical quality of the barriers between worlds, inviting the viewer to take sides. Absolute truth exists in *La ricotta*, and it is the truth of humanity's innate condition. For Capossela instead the postmodern crisis forces the individual to create a unique labyrinth of meaning, which starts and ends with the self. This is what Capossela means when he states in *Ovunque proteggi* that he is his own Minotaur: he is his own center. Piro says that this is a labyrinth in which no single pathway is superior to any other, where each point is infinitely connected to all others with no inside or outside (Piro, 2010:72). The track 'Brucia Troia' (Troy burns) references explicitly the end of the paradigmatic city, the end of fixed space: it inaugurates a place that can be anyplace. Individual wanderings, real or imagined, create place. This is a consciousness-raising effort, promoting a radically individualistic vision shared as a process, never as an outcome.

Capossela's postmodern reading of Pasolini revolves around an entirely new conception of the world. Yet Vinicio still returns occasionally to Pasolini's fixed geography, to the periphery as a radical break from the center, a primitive world within a modern setting. Capossela uses the metaphor of railroad tracks to symbolize radical alterity with a fixed geographical character. Consider this passage from Capossela's book *Non si muore tutte le mattine*.

> The neighborhood was situated behind the train station. Separated from the city by the railway. Only a high-speed freeway connected the two parts. The new modern complex built above the station, and other high rises with aerial antennas, as well as basement-level constructions, kept this neighborhood hidden from view.
>
> (Capossela, 2004:26)

Like Pasolini, Capossela is drawn to the margins in the effort to attain a sense of humanity. In *Non si muore tutte le mattine* he provides exact descriptions of people and places to capture the trivial minutiae of everyday existence. In a fictional city described as the barrio Capossela establishes narrative mood by describing spatial experience.

> Thus I returned to the streets of the barrio. I went to have a coffee in a bar run by Arabs, amongst tall blackened buildings, fruit and vegetable shops, and women porters who managed the buildings. I admired the passage of the morning from behind the café window. The custodians of the buildings threw water into the middle of the streets to rinse off their street doors. The tire sellers heaped up their wares on the tarmac. It was an experience to witness the arrival of merchants in the smoke and mist. From refrigerated trucks the carcasses of slaughtered animals were hauled out, fish carts made their appearance, as did braided garlic,

smoked cheese, cured meats, bags of pasta and flour, demijohns of oil and wine.

(Capossela, 2004:31)

For both artists the body is a privileged human site. Pasolini has a passion for the Eucharist because it expresses through the human body the idea of sin and redemption. Capossela is also fascinated by the Eucharist, an obsession which winds its way throughout the CD *Ovunque proteggi*, with most notable irony in the song the 'Il rosario della carne' (The rosary of the flesh). The sounds and words are a parody of the Eucharistic sacrament, recording the sound of a buzzing fly infesting deteriorating flesh. It is a hymn to the sacrality of the flesh.

> Flesh...
> Consolation of my flesh
> In the flesh that you are
> In the flesh you will return to
> Solitude of the flesh
> Of the souls in all flesh

(Capossela, 2006)

Body and spirit are one. The song uses a litany of terms to characterize the flesh: suffering, sacred, compassion, mercy, mourning, darkness, passion, penitence, ecstasy, chaos, scandal, sacrifice, decay, consolation. It ends with the idea of eternal life.

> It is not dead
> You are not dead
> In the flesh

(Capossela, 2006)

This parody of the Eucharist is also a philosophy of life. Flesh anchors the individual to a physical location within an ever-fluctuating landscape. We see this in the song about the Dead Christ procession, called 'L'uomo vivo' (Living man). The musical accompaniment is in the pompous style of marching bands in saintly festivals all over the Italian south. Exactly as in *La ricotta*, Christ is sacred as are all humans in a state of nature.

> He has left the Calvary and the shroud
> He has left the cross and his suffering
> He has risen from his slumber
> And now for always
> For always he is with us!

Christ belongs to the people who rescue him from the suffering to which he has been submitted by God the Father; he is the Christ of everyday affliction.

> If God the Father has abandoned him
> Now the townsmen escort him
> What a great feast to embrace him
> What a great feast to dine together

Breaking bread is the symbol of the Eucharist, and feasting together has a sacred character. God the Father has abandoned His own Son, and with Him all of us. The people rescue the Son according to their own sacramental procedures, hoisting the wooden Christ high up into the sky.

> Youth, let us hoist him up
> Let us cast him up
> Until he reaches high into the sky
> Until he can see the sea
> Until he can see the beauty which is life
> Which should never end

The Christ carried in procession rocks and dances with the crowd. At the end of the song, greeted with a round of applause, the wooden Christ sheds his pretense of divinity and goes back to being an ordinary man.

Capossela's resistance is directed against a postmodern urban condition that was theorized in the late millennial years. This postmodern condition is the outcome of forces operating on two different levels: one shatters spatial reference points, the other splinters the human subject. The contemporary market system undermines timeless features of humanity, cultivating an industry of desire which serves the needs of the market rather than the historically, socially and spatially integrated human subject. It creates a blank slate on which to inscribe paradigms beneficial to its pursuits, separating the individual from the moorings of shared experience, and eroding the community-supporting foundations of place. This reading of the postmodern world is not the elucubration of a political extremist, it is part of a critique of postmodern urbanism formulated in established academic circles. Postmodern urbanism was born in LA, and its most important theorist is Ed Soja. Soja theorizes the rise of the exopolis, the exemplar being precisely LA: 'This resplendent bazaar of repackaged times and spaces allows all that is contemporary (including histories and geographies) to be encountered and consumed with an almost Edenic simultaneity' (Soja, 1996:238). Soja is no defender of these urban trends, and is indeed the most important early advocate of greater consideration for spatial ontology in the social sciences. His description of the postmodern city could well be that to which Capossela responds.

Pasolini also believes that the source for the fragmentation of the world is the unbridled quest for market growth. Here too we have affinity with an extensive academic literature, especially the insights formulated by Baudrillard (1972) in the 1970s concerning the rise of new market systems based not on commodities but on the industry of desire. Holsteinization is an ironic way of describing the extreme form of this industry, achieved by corroding conventional community reference points to pave the way for the colonization of the individual. Again, the academic literature on the post-modern city provides insights which support this critique of contemporary urban life. According to one academic view, growth in the postmodern city is a 'quasi-random' phenomenon having nothing to do with the human subjects who inhabit these spaces.

> *Holsteinization* is the process of monoculturing people as consumers so as to facilitate the harvesting of desires, including the decomposition of communities into isolated family units and individuals in order to supplant social networks of mutual support with consumersheds of dependent customers. Resistance is discouraged by means of *praedatorianism*, i.e., the forceful interdiction by a praedatorian guard with varying degrees of legitimacy.
>
> (Dear and Flusty, 2000:61)

This description is the specter which haunts the works of Vinicio Capossela. Fortunately, it is not the city most Romans would recognize as theirs. Nor is it the city known to Pier Paolo Pasolini. It is an apocalyptic dystopia, the seeds of which Pasolini decried in his writings of the 1960s. Capossela represents a plausible extension of Pasolini's thinking in a contemporary world projected toward postmodernity. Yet the evidence suggests that Rome's community is hardly so fragmented as that described by the LA school. Indeed, resistance is situated in a conventional community where the rational and irrational are fused together in an ongoing reflection concerning the trajectories of modern growth.

Persisting spatial sacrality

Attention to the unmediated experience of place is part of Pasolini's legacy. Yet the ontological centrality of space is hardly a poetic invention, and is indeed part of a shared community concern. Today that concern is lived as an act of resistance, especially in street art and certain types of music. For instance, this resistance is seen in rap produced on the Roman periphery by Brasca Records. This is an independent label with recording facilities in the Corviale housing project (Fig. 6.2). The founder has lived in Corviale his entire life, and cherishes his existence in this peripheral outpost. The performers that work with him are also residents of Corviale, or other peripheral parts of the city where Brasca Records enjoys a reputation for promoting music at the city's edge. One of the musical approaches adopted by Brasca is remarkably similar to the stornelli described in previous pages. In this case the production of a rap song involves an electronically generated musical sequence, called the beat, over which the vocalist records spoken and sung words of their choosing. The producer generates the beat, and is sensitive to

Figure 6.2 Corviale. (Photograph taken in August 2020 © Gregory Smith.)

the potential lack of unity between words and sounds that Pasolini believed was a weakness of commercial pop music. He overcomes this potential risk by having the singer experiment with various vocal solutions until they find a style which fits the beat. There are no written-down lyrics, and the words can only be transcribed for a given recording or performance, in much the same manner of stornelli.

Some recordings, however, alternate improvised lines with written-down lyrics. This is the case of a tune recorded in 2013 and released as a video clip on Youtube the same year.[4] The recording is entitled 'Dove restiamo' (Where we are staying). The video clip shows images of the kilometer-long building, and in particular the skywalk that defines access to Corviale (Fig. 6.3). A group of rappers are moving along the skywalk, singing to the beat a written-down refrain. Among the various stylized hand gestures used by the group of vocalists is a downward movement indicating the space they are walking through.

> We are here and are going to stay
> Come and get us
> We won't stop anyway
> There are lots of us with a powerful spirit
> All of us with no regrets
> Come here
> We'll defend ourselves
> You want to attack us?
> But you are not ready

Figure 6.3 Corviale skywalk. (Photograph taken in August 2020 © Gregory Smith.)

> What are you going to do?
> We are the dim lights
> Burned out brains
> Are our fuel

While there is no direct connection with Pasolini, the artists involved in this recording know Pasolini as an advocate of the same resistance they themselves cultivate, defending the autonomy of the periphery. The song was recorded shortly after Rome's right-wing mayor of the time proposed that Corviale be torn down, claiming it was a symbol of peripheral decay. These citizens had a different idea.

Under the heading of urban arts, I include any creative endeavor which celebrates the spontaneous living city, including the visual arts, music, poetry and literature. I also include certain comics, which in Italy are an important literary form, and not only belonging to the superhero genre discussed by Umberto Eco (1976). Some comics have a critical approach to their portrayal of the city. This is the case of comics created by an artist who goes by the name Zerocalcare, an author born in the north of Italy but who grew up in Rebibbia, the neighborhood where Pasolini first encountered the periphery. Zerocalcare has received much public attention in Rome, and continues his impressive literary production, with some ten volumes of comics published since his first book came out in 2011. One critic writes of the 'Zerocalcare generation,' young people who grew up in the 1980s and 1990s in the urban periphery with few prospects for permanent employment, and a history of activism tied to social centers and emerging musical trends (Glioti, 2018). His first book is entitled *La profezia dell'armadillo* (*The prophecy of the armadillo*), initially self-published and then picked up by BAO Publishing which specializes in comics. It is written in a magical realist style strongly colored by its setting in the Roman periphery, of which Rebibbia is an especially powerful symbol. Though treated in an ironic way, the themes explored are those to which considerable leftist activism is devoted, especially the marginalization of Italian youth in terms of economic opportunities, the repression of political activism and the declining quality of urban life. His modern story-telling approach builds on the essentialized image of the city, implicitly referencing the spatial expression of an excluded community. It is a semi-autobiographical account starting with the author's own concern about future prospects. Rebibbia is no longer the borgata described by Pasolini, but is instead a modern characterless periphery which mixes together citizens from a variety of backgrounds all united by their exclusion from any possible privilege.

Celebrating the Roman periphery is today fashionable. This is particularly visible in street art, which is prominent in many parts of the periphery. An American artist who has made important contributions to Roman street art, Gaia, once wrote that street art should be illegal to be truly authentic (Gaia, 2013). He notes that requesting permission to produce street art exposes the artistic gesture to all kinds of curatorial and censorial procedures. The illegal act instead speaks directly to the viewer. He has a number of technically illegal art pieces in Rome, although some of his more recent art production has been commissioned. The commissioned pieces include his contribution to the Tor Marancia 'Condominium Museum' inaugurated in 2013, a publicly-funded project aiming to give visibility to one of Rome's historic borgate (Fig. 6.4). Tor Marancia has a drab reputation, even though it houses a community whose vitality is easily witnessed in the vibrant use of the borgata's poorly developed public spaces. In 2013 Rome elected a new progressive mayor who had spent many years in Philadelphia, and believed that public art could uplift a depressed periphery. Tor Marancia was particularly suited to the initiative, with its eleven buildings each four stories high, architecturally inconspicuous and painted with the shabby brown color used to decorate low quality public housing all over the city. Many urban activists decried the initiative as an expression of neoliberal tactical urbanism, painting over cracks created by the failures of the welfare state without addressing the real causes. The curator of the project was Galleria 999 which had already gained experience in this type of public art in the Ostiense underpass (Antonelli et al., 2015). It was intended as a participatory project where artists were selected from all over the world, and invited to paint walls facing mostly on the interior spaces of the housing project. The borgata covers an area of about ten hectares, with landscape maintenance provided almost exclusively by the residents. The participatory dimension was ensured by having each artist spend a week in the project to

Figure 6.4 Gaia's mural at Tor Marancia. (Photograph taken in April 2020 © Gregory Smith.)

interact with residents. Each artist presented three proposals to local citizens, and the final choice was made by residents whose windows look out onto the wall where the art piece was to be realized.

In local parlance Tor Marancia is dubbed Shanghai because of floods from the 1950s, the same years in which the flooding of the Yangtze River received much international press attention. In his piece Gaia includes the image of a fish to recall the floods, in local accounts said to be so severe that they could fish on the grounds of the borgata. The piece also has an image of the same building on which the mural is painted, in a technique we have already seen in a different connection. Another image is an orange soaring

into the sky like a balloon, a pun on the word Tor Marancia, where aran-
cia in Italian means orange, even though this has nothing to do with the
toponym's etymology. A final detail in this piece is the head of a classically
sculpted athlete, in reference to the temporary presence here of the crafts-
men who built the Foro Italico in the 1930s, a sports arena famous for its
statues. To make the piece even more place specific, the head is painted over
the window of a room which belongs to a local citizen who is a professional
boxer. All of the art pieces at Tor Marancia are linked to local curiosities
which celebrate place as well as the local community.

Gaia is a sophisticated artist and urban activist. He supports street art
because it expresses the idea of a living community: 'street art is not beautiful for what it produces, but for its potential [....] writing in the streets is
beautiful because something more will be done' (Gaia, 2013:2). It is a kind
of organic placeholder, an invitation for further engagement. Street art
pushes back against neoliberal urbanism, where livable housing stock is left
to degenerate by patient capital waiting for new investments. Yet there is a
danger, for street art can become a form of commercial place making rather
than an act of subversion. Developers market new urban frontiers by transforming authentic areas into sites of capital investment. Even Ostiense, once
a laboratory for innovative urban process, is now rated as one of the 'coolist
neighborhoods of Europe.'[5] The moral geography of the city has shifted,
and areas which were once beyond the pale of respectability have captured
the interest of the creative class, driving an inevitable process of gentrification (McAuliffe, 2012). The area that has been most affected by this process
of gentrification is Pigneto, the location where Pasolini filmed *Accattone*. A
commercial branding process has robbed this neighborhood of much of its
authenticity (Annunziata, 2014).

Pasolini is the symbol of resistance against the commodification of place,
and his image is visible in artworks distributed throughout the city. The
most significant mainstream art piece is no doubt William Kentridge's
reverse mural realized on the Tiber embankment walls in 2015, part of a
set of scenes celebrating Roman history called *The triumphs and laments*.
Eighty images are etched into the soot that covers the embankment wall,
one taken from the photograph of Pasolini's dead body after he had been
murdered in 1975.[6] Another image is contained in the theater called Teatro
India, dedicated to Pasolini by the leftist administration in 2000. Various
art pieces sprang up in Rome at the commemoration of the anniversary of
his death in 2015. Most poignant is that by the French street artist Ernest
Pignon, a stencil entitled *La Pietà*, where Pasolini is shown holding his
own dead body, reference to his isolation in both life and death (Fig. 6.5).
An Italian artist from Verona, now based in LA, Nicola Verlato, created
a mural in Torpignattara that interprets Pasolini's apotheosis (Fig. 6.6).
Rather than use the conventional iconography for an apotheosis, Pasolini
is brought down to earth where he joins the company of his mother, Virgil,
Petrarch and Ezra Pound. The most recent mural dedicated to Pasolini

Figure 6.5 Pasolini La Pietà by Ernest Pignon. (Photograph taken in August 2020
      © Gregory Smith.)

today is by Omino 71 and Mr Klevra on the façade of a school at Ostia, not
far from where Pasolini was murdered.[7] The list goes on, and includes many
locations outside of Rome, such as Scampia in Naples, the housing project
made famous through many mafia films (Fig. 6.7). There too his image is a
symbol of dignity and resistance against the forces of exclusion.

An unusual form of street art in Rome is produced by the Poeti anonimi
del Trullo (the anonymous poets of Trullo). They are well known for their
effort to revitalize in a subversive way Rome's most notorious borgata.
They started as a group of seven poets in 2010, styling themselves as the

Figure 6.6 Hostia (The Apotheosis of Pasolini) by Nicola Verlato. (Photograph taken in August 2020 © Gregory Smith.)

metropolitan romantic movement. Their poems were born on the walls of the city and on Internet, but are also published in a book of poems and short stories. The introduction to the book contains a manifesto comprising seven points (Poeti der Trullo, 2019). Point number one: 'Metro Romanticism is a poetic grassroots movement, starting in the neighborhood [quartiere] among the people, from the simplicity and complexity of everyday life.' Among the other salient points is number seven: 'Metro Romanticism travels in two different media, Internet and the walls of the city. It belongs to the street, and the city conceived as an immense blank sheet on which to write poetry.' They turn poetry into a democratic and revolutionary act, reaching out to people in their ordinary daily activities. The introduction states that pure and spontaneous sentiments prevail over reason. One poem makes

Figure 6.7 Pasolini in an image from Scampia, Naples. (Photograph taken in March 2019 © Gilda Berruti.)

implicit reference to Pasolini, a poem entitled 'Montecucco,' referencing the hill which oversees the valley where Pasolini filmed *Uccellacci e uccellini*. On the main street of the borgata one also finds a mural which reproduces a scene from this film. In another poem the author asks if their work will offend such historic poets as Leopardi and Pasolini. The poem 'Street art' notes that street art was born as an instrument of struggle against despotism. Like Pasolini, this poetry pushes back against the utilitarianism of the modern world. One poem can exemplify the effort.

Prima che er cervello
te lasci e scappi via

Figure 6.8 Mural from Trullo showing a handshake. (Photograph taken in August 2020 © Gregory Smith.)

> spegni la tv
> e leggi 'na poesia!
> [Before your brain
> up and runs away
> turn off the TV
> and read a poem!]
>
> (Poeti der Trullo, 2019:278)

In some cases murals are located alongside painted poems. This is the case of a mural showing a handshake united by a dove, with above written the words 'Da noi è così' (This is the way we do it). It is signed Diego (Fig. 6.8). To the right of the mural is a short poem signed 'Er Pinto Pdt,' where Er Pinto is one of the autonomous poets, and Pdt stands for 'Poeti der Trullo.' The poem says that the author has lived in the Trullo his entire life. He mentions the poetry of the neighborhood, and says the people here are true. It is a strong statement of geographically referenced authenticity.

> Ce vivo da sempre da quanno so' nato
> Nessuno fin'ora ve l'ha raccontato
> De la poesia che c'ha sto quartiere
> De quanto 'e persone ar Trullo so' vère
> [I have lived here since I was born
> Nobody has ever told you
> Of this neighborhood's poetry
> Of how the people of Trullo are real]
> Er Pinto Pdt

All of these art forms express commitment to the city as a shared space, a community with strong identity and solid local roots. Shared memory constitutes a force for resistance against trends leading to Ed Soja's exopolis. This is the foundation for principles of solidarity and moral growth. In Pasolini's terms these actions express commitment to progress rather than development. They also reference the separation between ordinary people and the dominant political-economic system. This spirit of cleavage has Gramscian roots and is given explicit support in such underground rap tunes as 'Curre curre guagliò' (Run run boy), which incites citizens to stand up against global capitalism. This famous tune dates to 1993, produced in social centers of Naples; today it is a Tarantamuffin classic (Wright, 2000). All of these activities express the importance of the community as a rallying point against exogenous forces of change, showing the continued relevance of the community as the foundation of what may be termed subaltern urbanism.

Notes

1 Hip Hop culture in the USA has been defined historically by rap, graffiti art and break dancing (Cornish, 2009). We have a similar interconnection among these art forms in Rome, although break dancing is rare.
2 https://www.famigliacristiana.it/articolo/vinicio-capossela-cosi-enrique-ira-zoqui-e-diventato-di-nuovo-gesu.aspx (Accessed October 15, 2020).
3 https://grazianograziani.wordpress.com/2006/02/27/parole-di-carne/ (Accessed August 13, 2020).
4 https://www.youtube.com/watch?v=YHb514gxfLM&list=PLDVy-i5E0x_O5YegAzHzBDW4X8TuYkj7M&index=3&t=0s (Accessed June 1, 2020).
5 '10 of the coolest neighbourhoods in Europe,' https://www.theguardian.com/travel/2020/feb/08/10-of-the-coolest-neighbourhoods-in-europe-paris-berlin-rome (Accessed June 3, 2020).
6 http://tevereterno.org/progetti/triumphs-and-laments/ (Accessed August 15, 2020).
7 https://www.artribune.com/tribnews/2015/05/pasolini-roma-e-la-street-art-la-formula-che-funziona-nuovo-murale-a-ostia-firmato-da-mr-klevra-e-omino71/ (Accessed August 15, 2020).

Bibliography

Annunziata, Sandra (2014) 'Gentrification and public policies in Italy,' *The changing Italian cities: emerging imbalances and conflicts, L'Aquila, GSSI Urban Studies-Working Papers*, 6:23–34.
Antonelli, Stefano, Francesca Mezzano and Gianluca Marziani (2015) *Big city life. Tor Marancia*. Rome: Castelvecchi Editori.
Banda, Alessandro (1990) 'Appunti sul Leopardismo di P. P. Pasolini,' *Studi Novecenteschi*, 17:39:171–195.
Baudrillard, Jean (1972) *Pour une critique de l'économie politique du signe*. Paris: Gallimard.
Borgna, Gianni (1992) *Storia della canzone italiana*. Milan: Oscar Mondadori.

Capossela, Vinicio (1996) *Il ballo di San Vito [CD]*. Milan: CDG East West.
Capossela, Vinicio (dir.) (1998) *Live in Volvo*. Milan: CDG East West.
Capossela, Vinicio (2000) *Canzoni a manovella [CD]*. Los Angeles: Warner Music.
Capossela, Vinicio (2004) *Non si muore tutte le mattine*. Milan: Feltrinelli.
Capossela, Vinicio (2006) *Ovunque proteggi [CD]*. Los Angeles: Warner Music.
Capossela, Vinicio (2012) *Rebetiko Gymnastas [CD]*. Los Angeles: Warner Music.
Capossela, Vinicio (2013) *Tefteri*. Milan: Il Saggiatore.
Carreri, Francesco (2006) *Walkscapes, camminare come pratica estetica*. Turin: Picccola Biblioteca Einaudi.
Certeau, Michel de (1990) *L'Invention du quotidien, i. Arts de faire*. Paris: Gallimard.
Conti Calabrese, Giuseppe (1994) *Pasolini e il sacro*. Milan: Jaca Book.
Cornish, Melanie J. (2009) *The history of hip hop*. New York: Crabtree Publications.
CSOA Forte Prenestino (2016) *Fortopìa - storie d'amore e di autogestione*. Rome: Fortepressa.
Dear, Michael J. and Steven Flusty (2000) 'Postmodern urbanism,' *Annals of the Association of American Geographers*, 90:1:50–72.
Dei, Fabio (2013) 'Dal popolare al populismo: ascesa e declino degli studi demologici in Italia,' *Meridiana*, 77:83–100.
De Martino, Ernesto (1961) *La terra del rimorso. Contributo a una storia religiosa del sud*. Milan: Il Saggiatore.
De Mauro, Tullio (1999) 'Pasolini's linguistics,' in Baranski, Zygmunt G. (ed.) *Pasolini old and new: surveys and studies*, pp. 77–90. Dublin: Four Courts Press Ltd.
Eco, Umberto (1976) *Il superuomo di massa. Retorica e ideologia nel romanzo popolare*. Milan: Bompiani.
Gaia (2013) *Second cities*. Self-published Zine. (Available at https://gaiastreetart.com/Accessed August 5, 2020).
Giardinazzo, Francesco (2009) *La culla di Dioniso. Storie musicali del passato prossimo*. Genoa: Marietti.
Glioti, Oscar (2018) 'Generazione Zerocalcare,' *L'Espresso*, 5 November.
Grazioli, Margherita (2017) 'From citizens to citadins? Rethinking right to the city inside housing squats in Rome, Italy,' *Citizenship Studies*, 21:4:393–408.
Hennessy, Jeffrey J. (2015) 'Deterritorialization and reterritorialization,' in *Atlantic Canadian Popular Music*,' *MUSICultures*, 42:1.
Lamanna, Gaetano (2019) 'L'Occupazione di case somiglia alla lotta contro il latifondo,' *Il Manifesto*, 18 May.
Lefebvre, Henri (2019) *Rhythmanalysis: space, time and everyday life*. London: Bloomsbury.
Lynch, Kevin (1960) *The image of the city*, Cambridge, MA: MIT Press.
Lyotard, Jean-Francois (1979) *The postmodern condition*, Manchester: Manchester University Press.
Mancusi, Aldo Luigi (2012) 'Vinicio Capossela, La Ballata di Caronte,' http://brutalcrush.com/2012/06/07 (Accessed June 30, 2014).
McAuliffe, Cameron (2012) 'Graffiti or street art? Negotiating the moral geographies of the creative city,' *Journal of Urban Affairs*, 34:2:189–206.
Mitchell, Tony (1996) *Popular music and local identity: rock, pop, and rap in Europe and Oceania*. Leicester: Leicester University Press.
Montagna, Nicola (2006) 'The de-commodification of urban space and the occupied social centres in Italy,' *City*, 10:3:295–304.

Mudu, Pierpaolo (2014) 'Where is culture in Rome? Self-managed social centers and the right to urban space,' in Marinaro, Isabella Clough and Bjørn Thomassen (eds.) *Global Rome: changing faces of the eternal city*, pp. 246–264. Bloomington: Indiana University Press.

Pasolini, Pier Paolo (1950) *Roma 1950: diario*. Privately printed manuscript.

Pasolini, Pier Paolo (1955) *Canzoniere italiano. Antologia della poesia popolare*. Parma: Guanda.

Pasolini, Pier Paolo (dir.) (1964) *Il Vangelo secondo Matteo*. Rome: Arco Film.

Pasolini, Pier Paolo (1976 [1957]) *Le ceneri di Gramsci*. Milan: Garzanti.

Pasolini, Pier Paolo (2005) *Heretical empiricism*. Translated by Ben Lawton and Louise K Barnett. Washington, D.C.: New Academia Publishing.

Pasolini, Pier Paolo (2006) *Accattone, Mamma Roma, Ostia*. Milan: Garzanti.

Pasolini, Pier Paolo (2009 [1960]) *Passione e ideologia*. Milan: Garzanti.

Piro, Loredana (2010) *Vinicio Capossela. Ri-cognizione geografica di una flanerie*. Milan: Mimesi Edizioni.

Plastino, Goffredo (2008) *'Introduzione'* to Alan Lomax, *L'anno più felice della mia vita. Un viaggio in Italia 1954-1955*. Milan: Saggiatore.

Poeti der Trullo (2019) *Poeti der Trullo*. Rome: Associazione Culturale Metroromantici.

Salis, Stefano (2013) 'Quaderno di conti e di canti,' *Il Sole 24 Ore*, 23 June.

Savona, Virgilio and Michele Straniero (1994) *Canzoni italiane*. Milan: Fabbri Editori.

Shapiro, Susan O. (2013) 'Pasolini's Medea: a twentieth-century tragedy,' in Nikoloutsos, Konstantinos P. (ed.) *Ancient Greek women in film*, pp. 95–116. Oxford: Oxford University Press.

Soja, Edward W. (1996) *Third space. Journeys to Los Angeles and other real-and-imagined places*. Oxford: Blackwell Publishing.

Tommasini, Alessio (2019) 'Street art e situazionismo. Intervista a Hogre,' *Artribune*, 29 January.

Wright, Steve (2000) '"A love born of hate" autonomist rap in Italy,' *Theory, Culture & Society*, 17:3:117–135.

7   Contemporary relevance

Development or progress

Here we return to Pasolini's contrast between development and progress. Development is what industrialists want, a way to produce and market superfluous commodities; it is a right-wing concept (Pasolini, 2018:175). Progress is instead a progressive social and political notion. Support for progress understood in this way has been growing over the decades, and seems ever more relevant in a post-Covid-19 world. It is a contemporary expression of the modern ecological paradigm. Consider the criticism of the massive efforts to reconstruct the Italian economy at the time of this writing, efforts that at times seem oriented to re-establish the status quo ante Covid-19. Many voices urge for a new economic model predicated on principles of social and environmental sustainability. They do not reject the market as a vehicle for advancing societal growth, but advocate healthier conditions for the environment, for workers and the economy itself (Pileri, 2020).

The year 1989 is a symbolic watershed in modern European history, opening the way to a post-ideological age. Before that year membership of the communist party could provide a shorthand definition of progressive identity (Bobbio, 1994). Pasolini, of course, cultivated his progressive image by defining himself as a Marxist and a communist. Yet we know that critics in his own day disputed the claim, and even writers today insist that Pasolini was essentially a reactionary. This is the position taken by the prominent economist Giulio Sapelli (2015) in a book written about Pasolini's political economy. In Sapelli's view, Pasolini's most backward position is precisely the belief that economic well-being destroys some inherently sacred human quality, including the spontaneous character of congregation (Sapelli, 2015:187). This is Pasolini's 'reality,' which Sapelli rejects as belonging exclusively to the realm of the imagination. Sapelli martials a surreptitious social nominalism against the peasant's socially grounded idea of human perfection which so fascinated Pasolini, and with it the problem of sin and salvation resolved within the human congregation. For Pasolini, consumer culture obliterates the distinction between good and evil, and so destroys the moral fabric of society. In the interpretation advanced by Sapelli, Pasolini's wish to suppress the 'anguish' of consumer society preserves the conservative system of social relations guaranteed by integration in a circumscribed community (Sapelli, 2015:23). Such integration is aligned with the 'unhappy happiness' we witness in so many of Pasolini's films and novels (Sapelli, 2015:160). This fascination for the morality of conservative peasant society allows Sapelli to place Pasolini in the class of the 'great reactionary poets,' in the same league as the fascist Ezra Pound (Sapelli, 2015:152).

Today identification as a Marxist is no guarantee of a progressive stance. The PCI disappeared after 1989, and since 2008 no party in parliament has supported a Marxist agenda. Yet there are many progressive movements in Italy, and they are often unaligned to political parties. They are part of elastic networks, often operating within an environmental paradigm. One of the best-known intellectuals falling within this general paradigm today is the French anthropologist Serge Latouche. He is most noted for his theory about happy degrowth (Latouche, 2015). Latouche writes that we must abandon the economic system which reduces society to an instrument of production, treating human beings as waste products. It is a system which has 'disastrous implications for the environment, and therefore for humanity' (Latouche, 2015:8). His three concerns are the environment, humanity and society. These are cardinal points for many progressive movements, although generally outside a formal political framework. We are not far from Pasolini's stress on progress over development. Indeed, Italy's degrowth movement flourishes, headed up by an energetic activist who lives in the western Roman periphery. She is strongly engaged in her local neighborhood, and has high visibility on a national scale. She pushes for new economic models based on social solidarity with deep community roots.[1] At a national level, Serge Latouche's ideas were embraced by the founder of Italy's Five Star Movement which began to enjoy electoral success in 2013.[2] While neither the happy degrowth supporters nor the Five Star Movement advocates a return to peasant communitarianism, they encourage sensitivity to contextual factors which repel the social nominalism of advanced consumer society. In the sense described here, they are on the side of progress over development.

The progressive movements broadly associated with a principle of degrowth represent significant variety drawn together in a wide range of networks. A key focus is quality of life, with strong attention to environmental factors. Within this activism we find, for instance, the CSA – Comunità di sostegno all'agricoltura (community for sustaining agriculture) which aims to create more support for small-scale urban agriculture by forging direct links between producers and final consumers. The two interact in all phases of the production process, working to reduce the mediation of an impersonal monetary market principle.[3] We also find an abundance of citizen farms on the Roman periphery, and successful young cooperatives like the Cooperativa Coraggio which operates on the city's northern fringe.[4] This cooperative, made up of young people from the Roman periphery, champions a new political, cultural and economic model, organizing gainful activities with strong attention to factors that go beyond economics. They are young people in their thirties, who like many Romans have long been employed as precarious workers in diverse economic sectors. They see sustainable urban farming as providing a modicum of security, while allowing them to pursue political and social goals similar to those identified here. Openness to market forces tailored to their own vision makes them different from the Roman progressive movements of the 1990s, with their rigid Marxist ideology. The new activists accept the idea of a market harmonized with principles of sustainable human, social and environmental growth (Smith and Berruti, 2019). Like Serge Latouche, they advocate economic action which goes beyond the polarity between neoliberalism and Marxism, to embrace concern for a full range of contextual factors.

Linking these contemporary movements to the heritage of Pier Paolo Pasolini is no simple task, given the substantial historical distance between the situation today and the early postwar period. Yet for all the many differences, Latouche cites Pasolini as a precursor to the degrowth movement, an early critic who decried the cultural desolation and smothering political effects of rising consumer culture (Latouche, 2015:29). For both Latouche and Pasolini, the modern consumer is hetero-directed, lacking capacity for autonomous action: he is the average man who supports all the abuses Pasolini details in *La ricotta*. In this catalogue of abuse, we find political indifference, racism and exploitation - but no mention of environmental devastation. This is a major difference between today and years past when with few exceptions environmentalism was weak. Yet Latouche finds affinity with our Italian author owing to the dim view taken of an economic system driven by the senseless pursuit of infinite growth. For Pasolini Marxism was not the answer, at least not Russian style communism with its focus on industrial growth (Pasolini, 2018:176). Latouche echoes this sentiment.

While environmentalism is not part of his political agenda, Pasolini shows strong sensitivity to environmental factors broadly understood. This is seen in his celebration of the simple peasant life, whose cyclical character is endowed with a sacred quality. This is the life Pasolini cherished. When asked in an interview of 1970 whether he loved life, he responded: 'I love it fiercely, even desperately. I believe that this ferocity and this desperation will only lead to my destruction. I love the sun, the grass, youth.'[5] In *Il sogno del centauro* Pasolini states that since childhood he had an irresistible desire to admire nature and humankind (Pasolini, 1989:21). Nature and the physical world are an essential part of the universe Pasolini describes.

In order to give substance to the suggested historical parallel, Latouche asked the prominent historian Piero Bevilacqua to produce a study concerning the relevance of degrowth principles in Pasolini's writings. Bevilacqua (2014) accepted the challenge, and demonstrates admirably how Pasolini fits into this line of thought. Pasolini alone in his day speaks out against the celebrated fruits of material prosperity, taking on the established political forces of the left and the right which extol economic development as having the power to lift Italy out of its backward state. Few great intellectuals in modern Italian history have had the courage to stand up against the 19th-century idea of progress that is widely celebrated as the founding stone of national unification. Challenging this praise of uncritical economic growth is what unites Latouche and Pasolini.

One of Pasolini's most celebrated metaphors for the bitter fruits of modernity is his claim that the disappearance of fireflies marks a watershed in Italian history. Here too we see that while the environment is never the explicit object of Pasolini's attention, nature is never distant from his thought. Living in harmony with nature is the foundation of peasant society. His poems celebrate the natural environment he had known in his youth in Friuli, the rhythm of life organized around an ancestral cycle (Bevilacqua, 2014:27). This thinking resonates with many activists in Italy today, and not only young progressives. Many citizens embrace a 'back to nature attitude,' and all accept fireflies as an important indicator of environmental health.

There is inevitable difference between the celebration of nature by ordinary citizens, and the pursuits of committed urban activists. Today these pursuits often take the form of acknowledged affinity with Marxist activism of the 1990s and early millennial years, generally referred to as the movimento romano (Roman movement). The Roman movement has its roots in antagonistic activism dating to the 1960s, when the idea of revolution was taken out of the factories and put into the streets and squares of the city. This historic movement has a strong degrowth flavor, especially through the ideas of autoriduzione (self-reduction) and autoproduzione (self-production).

Class struggle beyond the factory

Many initiatives in Italy aspire to reduce the impact of the pure market economy. An important set of these initiatives involves Universal Basic Income (UBI). Italian UBI supporters wish to downsize the market so citizens can devote more time to non-commercial interests and civic engagement (Gobetti and Santini, 2018). These pursuits are the hallmarks of the moral progress we have been discussing. A variant of UBI was part of a national law passed at the behest of the Five Star Movement in 2019. The right to work is famously celebrated in Article 1 of the Italian constitution, which describes the nation as a democratic republic founded on work. Work has long been central to the thinking of both capitalists and workers; even peasant struggles historically focused on the creation of work through land occupations. UBI runs against this long-standing trend, expressing continuity with experiments promoted by the autonomist movements of the 1960s and 1970s which are today considered the foundation of Italian theory (Gentili and Stimilli, 2015). Pasolini finds some mention in the debate generated by these radical extra-parliamentary movements, but not much (e.g., Virno and Hardt, 1996). This omission is hardly surprising, considering Pasolini's description of the political unrest of 1968 as being a revolution of the bourgeoisie against the bourgeoisie, one generation against another, pitting figli di papà (Daddy's boys) against their privileged fathers (Pasolini, 2005:150). Yet though an outsider who never identified with any movement, Pasolini was still part of the intense political ferment that characterized Italy in those years. Italy then was a laboratory for experimentation with innovative political initiatives which have ripple effects down to the present day. The Italian '1968' was longer than in other European countries, stretching from the 1960s to beyond the activism of 1977. Indeed, 1968 was European, while 1977 was purely Italian (Nicolini, 2011).

In an innovative effort of the 1960s, political struggle was shifted beyond the economic system narrowly considered. For our concerns, the most important instance of this innovation was the connection of class struggle to spatial politics. It was in the light of this experimentation that some movements chose not to restrict political struggle to the factory precinct. The position was clearly stated by one of Italy's prominent extra-parliamentary groups in the early 1970s.

> Community struggle in Italy has gone beyond the trade-union tradition which limits the class struggle to the fight for higher wages. The Italian working class have recognized that their needs for a freer and happier life cannot be realized by increasing the spending power of individual groups of workers. Any gains made inside the factories have been countered by the bosses' use of inflation and property speculation.
> (Lotta Continua, 1973:79 – Preface by Ernest Dowson)

This statement captures the logic of non-work, urging workers to withdraw from exploitation, and engage in life-sustaining activities outside the realm of wage labor. This mass flight from capitalism also involved a strategy of autoriduzione (self-reduction), limiting the use of goods and services furnished by the capitalist market and state, and the allied concept of autoproduzione (self-production). This was a precursor to the logic of degrowth, considering that one of the most important features of the Roman degrowth movement today is self-production and non-monetary exchange. The abundant literature on self-production testifies to growing interest in techniques which elude consumer logic (e.g., Cuffaro, 2014). The touchstone of these historic movements is self-valorization, involving production outside capitalist circuits. In Rome one of the best examples of this principle today is the Communities for Sustaining Agriculture we have already mentioned.

The ideas expressed in those years aspired to forge a new concept of the city. The Italian aspiration went beyond Lefebvre's (1996) 'right to the city,' representing something closer to the concept of autogestion Lefebvre (2009) theorized in his later years. Autogestion describes a utopia set against spatial commodification, a more contemporary concept than the right to the

city, and one with special pertinence to Italy. The right to the city is ultimately based on private property and possessive individualism, and according to some authors movements promoting these rights may exacerbate the
uneven power relations they aim to overcome (Gray, 2018).

The battle cry launched by Lotta Continua in the 1960s was 'Take over
the city!' The right to the city was asserted not within property rights, but
against them. This was the meaning behind the autonomist movements,
whose symbol of the letter 'A' can still be seen scrawled on walls all over
the periphery. Between 1969 and 1975 twenty thousand habitations were
squatted, in addition to actions involving self-determined rent reduction,
especially in public housing (Cherki and Wieviorka, 2007:73). A new idea
of struggle arose at the juncture of the factory and the city. In this self-
reduction movement workers decided how much they wanted to work, and
how much they were prepared to pay for services. It was a form of direct
action, making no attempt to cooperate with the formal political system.
This movement had a strong interclass component, bringing together
diverse community segments coordinated by neighborhood associations
which have persisting importance today.

The early 1970s witnessed the displacement of mass workers, and their
decomposition as political subjects, accompanied by new forms of political
leadership. The world that emerged was characterized by far more nuanced
relationships than those imagined by the Marxist activists of the PCI.
The spatial dimension of this change was the more complex organization
of the city (Gray, 2018:325). Gone was Italy of the 1960s, when masses of
workers emigrated from the rural south to the industrial north where they
were segregated in intolerable conditions, with as many as eight workers
sleeping in a single room (Gray, 2018:327). The powerful social cleavages
of this earlier period were documented by Pasolini in Rome's peculiar cultural and political context. The subsequent transformation led to new waves
of community activism, especially in the 1990s, when movements reacted
to uncontrolled urban growth by occupying parts of the city which they
could give over to new social functions. It was a total political struggle on
a diminished scale. Street art and Italian rap are rooted in this activism,
coordinated by the social centers which sprang up all over Italy, known by
the acronym CSOA, or Centro Sociale Occupato e Autogestito (Occupied
and Self-Managed Social Center). This was a politically antagonistic escape
into pure self-determination.

The politically antagonistic movements described here all have strong
community roots. As I have noted repeatedly, neighborhoods and neighborhood associations still have critical importance in the life of the city
(Della Porta, 2004). The fostering of strong local communities especially by
progressive political movements raises an important theoretical question.
Bevilacqua reminds us that Adorno and Horkheimer criticized Durkheim's
praise of community-based social solidarity as representing a suffocating
form of domination, coercion imposed from above (Bevilacqua, 2014:26).

According to this interpretation, such solidarity should be shunned in a liberated society. So how can progressive forces in Italy, from Pasolini to the autonomist movements, and down to contemporary times, advocate defense of the community? We started our study by recalling Bruno Latour's treatment of Durkheim's community, inviting us to see this as a process rather than an established fact. This is Pasolini's reality, which Sapelli dismisses as pure imagination. But it is precisely imagination that can provide a flexible definition of the modern community. Certainly, none of the progressive movements mentioned here embrace conservative peasant society. They embrace a geographically-circumscribed community articulated within a broader vision which is global in character.

The mobilization of citizens in neighborhood associations in Rome today cuts across class identity. This is unsurprising given the challenges faced by most city dwellers – and not only the poor. Inept politics and widespread corruption ensure mounting challenges for the quality of urban life (Cellamare, 2018). Added to the decline in quality of life is the realistic perception of risks for downward mobility. Wide public awareness circulates concerning the labile conditions of the Italian middle class, and the likelihood of joining the ranks of the downwardly mobile (Ventura, 2017). If the leisure society of mass consumption fails, one can always find comfort in spontaneous urban pleasures. The void of effective government strengthens community associations all over Rome (Celata et al., 2018). Indeed, the periphery seems to have a greater abundance of neighborhood associations than the center. This is perhaps because life outside the center is less strictly policed, and affords greater opportunities for citizen-directed initiatives. The vibrancy of life in the periphery is reflected in street art, citizen associations, comics and a budding literature such as Pontale's recent *La Roma di Pasolini* (Pontale, 2019). The periphery is the real Rome, in contrast with the spectacle of the historic center progressively deprived of its native residents (Herzfeld, 2009). Of course, all areas of Rome face major challenges today; but so do cities all over the world. Coppola's study of selected American cities, aptly entitled *Apocalypse town* (Coppola, 2012), was so well received in Italy because it shows the depths of urban decline in the world today. There is much debate on Rome's perceived decline as an urban environment, with an abundance of *declinisti* (declinists) who sound an alarm which is much like a call to arms (Mosco, 2018). The city's greatest hope to reverse decline is precisely the ideal city that people have in their heads, which they strive to realize even in the absence of promising physical conditions. The pioneer of that effort is certainly Pasolini.

Pasolini courted the contradiction expressed in his term sineciosi, and his poetry admits no single reading. But this is the nature of poetry: it raises provocative questions concerning challenges of the present without furnishing underwhelming answers. The contradictory readings of Pasolini leave one solid impression, and that is the importance of ordinary lives in ordinary places. Pasolini is today 'a haunting presence in Italy's cultural landscape'

(Bernini, 2015:3). His memory is writ big on the city of Rome, and indeed on the whole nation, both in its image of itself and in perceptions from abroad. How transferable lessons drawn from Rome may be to other national settings is an open question. But some evidence supports the idea that his effort to identify areas of underprivilege and advocate their rights within an increasingly divided global arena has significant transferable value.

Subaltern urbanism

The video installation *Devo partire. Domani* (*I must leave. Tomorrow*) directed by Ming Wong and hosted at Naples's modern art gallery (PAN) in 2010 is remarkable.[6] It is a remake of Pasolini's celebrated film *Teorema* (1968), originally shot in Milan as a piercing critique of bourgeois family life. The video instillation shows footage shot in Naples's industrial periphery, using Asian actors to reproduce scenes from the original film in new locations which have recognizable sematic affinity with the original settings. In the video installation, individual film scenes are looped simultaneously in physically separate viewing spaces, all visible from a single location. This technique elides the original plot sequence, creating a simultaneity which excises an important temporal dimension to allow readings from different cultural perspectives. Collapsing the plot structure through temporal elision, along with the recontextualization of location and character, transmits a powerful message of universality. The installation exposes the fecundity of images whose meanings are strangely familiar in an unfamiliar assemblage combining the intentions of an artist from Singapore with the vision of an acclaimed Italian filmmaker reworked in a foreign setting with actors from an entirely unexpected background.

Pasolini would no doubt have appreciated this installation, considering his fascination for the world outside of Italy. It also provides a measure of the fascination people from abroad have for Pasolini. His realism, and his support for untrammeled spontaneous life outside the bounds of the advanced market system, evidently have high capacity to transcend cultural barriers. We know that Pasolini rejected this classification, but most international viewers associate him with Italy's Neorealist movement which combines detailed documentation of real-life situations with a powerful critique of social injustice. Going back to Latour's claim that actants include structural traits, we might see reference to Pasolini in these international settings as urging a response to the global presence of injustice and exclusion (Latour, 2007:54). To pursue the Latourian reflection, we can note that recognition of this trait as an actant promotes and encourages an attitude of resistance. Alienation is also a trait, whose cultural transferability is finely demonstrated by *I must leave. Tomorrow*. In contexts such as these, Pasolini's message is an actant in the experience of the urban learning assemblage. The message is as dense as it is consolidated, achieving a concrete plasticity in the most disparate settings.

Being-in-the-world unfolds in a state of flux, a process of dwelling which draws from both imagination and life on the ground. It is primarily practical rather than cognitive (McFarlane, 2011:21). In this flux, the assemblage is a critical tool in describing unity across difference, and the forms of inter-action that constitute the world. Critical is the idea of the city itself, which too changes according to time and context. The assemblage is both an ori-entation to the world and the object of the world (McFarlane, 2011:23). The process of imagining is part of the experience of the world, supported by different ways of knowing a spatial context, including poetry and narrative. Cultures are relational, not bounded or territorial, filled with contradiction and tension. The contradictions involve awareness of power relations within society and among societies. In Italy we have described the contrast between north and south, between center and periphery. This is part of an internal history of domination and resistance, mirroring biases on a global scale. In both settings, national and global, local knowledge expresses contrapuntal awareness of the history of domination, intertwining histories as overlapping narratives (Said, 1994:18). In Italy Pasolini provides an implicit contrapuntal perspective, where the poor of the south and the periphery see themselves refracted through the eyes of privilege in a way that influences the vision of the poor as well as the dominant classes. This polarity of viewpoints was the-orized by Gramsci, and is the foundation for the study of subaltern cultures.

All of Pasolini's works operate within this contrapuntal framework. He essentializes and liberates, combining an awareness of spatial discredit with a desire to push back against such discredit: nothing could be more con-tradictory. He does this not only in relation to Italy's peasant tradition, or the Roman periphery, or the Italian south; he does this in his vision of the world. His gaze is deeply connected to alterity, subalternity and the idea of a pan-south (Trento, 2010). His interest in global oppression is evident especially in his mature period, with an early instance of this concern being found in the 1962 poem 'Prophecy,' usually known as 'Alì dagli occhi azzurri' (Alì with blue eyes).[7] The poem was dedicated to Sartre following conversations which inspired Pasolini to write on this topic. It was also inspired by Pasolini's encounter in a Roman cinema with his future part-ner of many years, Ninetto Davoli. Ninetto was an unschooled resident of borgata Gordiani who spoke to Pasolini about Persians amassed at the national borders, and the millions who had migrated to Rome now living at the end of the tramlines. He described the Persians as beautiful, and said their leader was Alì with Blue Eyes.

> Alì dagli Occhi Azzurri
> uno dei tanti figli di figli,
> scenderà da Algeri, su navi
> a vela e a remi. Saranno
> con lui migliaia di uomini
> coi corpicini e gli occhi

> di poveri cani dei padri
> sulle barche varate nei Regni della Fame. Porteranno con sé i bambini,
> e il pane e il formaggio, nelle carte gialle del Lunedì di Pasqua.
> Porteranno le nonne e gli asini, sulle triremi rubate ai porti coloniali.
> Sbarcheranno a Crotone o a Palmi,
> a milioni, vestiti di stracci
> asiatici, e di camice americane.
> Subito i Calabresi diranno,
> come da malandrini a malandrini:
> «Ecco i vecchi fratelli, coi figli e il pane e formaggio!»
> Alì with blue eyes
> one of the many children of the children
> who will travel in from Algiers, on boats
> with sails and oars. With him
> will come thousands of men
> with small bodies and the eyes
> of the poor dogs of their fathers
> on boats which embark from the Kingdom of Hunger. They will bring with
> them children,
> bread and cheese in the yellow paper of Easter Monday.
> They will bring their grandmothers and donkeys, on triremes stolen from
> colonial ports.
> They will beach at Crotone or at Palmi, millions of them, dressed in Asian
> rags and American shirts. The people of Calabria will say, rogue to rogue:
> 'Here are our elder brothers, with children, bread and cheese!']
>
> (Pasolini, 1976:97)

Pasolini's interest in Africa and Asia was cultivated through readings and travels, and expressed in writing and filmmaking (e.g., Pasolini, 1990). This interest also motivated his decision to help organize in southern Italy a film festival in 1959, called the Festival del Neorealismo, which gave priority to films made in the global south. Neorealism was more than an artistic movement, it was a political initiative. The supporters of the film festival believed that cinema could aid the oppressed, reinforcing solidarities between the experience of southern Italians and disenfranchised communities all over the world (Ruberto and Wilson, 2017:142). Cesare Zavattini, a leading theorist of Neorealism, argued that film should engage the audience ideologically, raise social consciousness, question hierarchies, and dismantle dominant structures. Neorealist films were always 'radical, experimental, and political forms of expression' (Ruberto and Wilson, 2017:140). Italian Neorealism in film took shape in the 1940s, and had immediate influence in many parts of the world, with filmmakers from the southern hemisphere operating under its inspiration. Some even studied and worked in Italy, such as the Moroccan filmmaker Souheil Ben-Barka, who studied at the Italian cinema academy and worked with several Italian directors, including Pasolini on the film *The Gospel according to Matthew*.[8] He was a leading light of African Neorealism, and went on to become the director of the Moroccan film institute. Another egregious example of this influence is the father of Egyptian Neorealism, Salah Abu Seif, who came into contact with Italian Neorealism during a trip to Italy in the late 1940s. He was impressed by the high artistic quality of the films being shot then in Rome, and the low cost. When he returned to Egypt, he directed several films using Neorealist film techniques and profound social commentary.[9] Linkage with African filmmakers continues into the present day, including the Algerian film-maker Rachid Benhadj, whose *For bread alone* (Benhadj, 2006) was funded by an Italian producer. The film draws from a novel written by Mohamed Choukri containing a description of life among the Moroccan underclass with strong stylistic affinity to Pasolini's writings.

Neorealism had important impact on many countries in the global south, including Latin America (Ruberto and Wilson, 2017). But here I would like to limit my attention to the Indian film director Satyajit Ray. He screened Italian Neorealist films in London for the first time in the early 1950s, and wrote of the experience saying he knew immediately that he would eventually use this technique for films set in India (Ruberto and Wilson, 2017:140). This influence is especially evident in the Apu Trilogy, with its powerful images of poverty and exclusion, and deep empathy for the poor. The films that influenced Ray predate Pasolini's filmmaking career, but the cultural matrix which shaped Pasolini, with its critical sensitivity to social justice, was the same that shaped the great Indian film director.

What is surprising about India is not only the impact of Italian Neorealism on film production, but also the use of Gramsci's theoretical model to explore Indian social issues. Italian Neorealism in literature has its roots in the 1930s, but its heyday was in film in the 1940s and 1950s, precisely when the works of Gramsci were first published in Italian. Thus, there is no direct link to Gramsci in the origins of Neorealism, but the discovery of Gramsci's reflections, especially on popular culture, coincided with the success of the genre's filmic expression. The topic is complex, but we must note that in postwar Italy the relationship between arts and politics operated under the counterhegemonic influence of the PCI. The PCI organized a cultural commission which brought together prominent intellectuals, as well as artists and filmmakers, and stressed the duty of communist artists to document the conditions of subaltern people bearing in mind the political function of these portrayals (Vittoria, 1990). Pasolini also operated under a Gramscian framework, but criticized the PCI's attempt to control progressive artistic movements as a form of Zdanovism, taking the term from Stalin's chief cultural advisor (Pasolini, 2005:159–163). Thus Neorealism was an artistic project, but also part of a political debate that aimed to raise social consciousness and dismantle dominant structures that effectively obscured knowledge and recognition of subaltern cultures thus impeding their advancement.

The relationship between politics and the arts was no doubt different in other national settings. But a constant feature of Neorealism was attention to the manner of portraying subaltern life in a way that bore some relationship to the writings of Antonio Gramsci. Thus, the style of investigating and portraying the life of ordinary people that Satyajit Ray brought to India in the 1950s had at least an indirect link to Italian reflections on subaltern people. I make this premise in relation to the emergence some years later of an explicitly Gramscian research project founded as the Indian Subaltern Studies Collective in the 1980s. The collective's knowledge of Gramsci actually came from a British source, namely the English language translation of the *Notebooks* published in 1971 (Macri, 2016:131). Their research inaugurated a method of investigation whose influence continues in the present day, using the term 'subaltern studies' to establish an explicit link with the great Italian Marxist writer.

It is perhaps inevitable that Italian scholars should claim that the Indian reading of Gramsci lacks philological rigor, yet their interpretations provide for penetrating analysis of Indian social and historical process (Macri, 2016:138). A type of manifesto for the group was published in 1982, where their founder, Ranajit Guha, describes a project drawing from the six points Gramsci outlined in the 'Notes on Italian history' (Gramsci, 1982:52). These points envision a systematic study of subaltern groups, starting with their objective formation, their affiliation with dominant groups, their autonomy as social formations and allied aspects. The manifesto was published in a volume containing studies on Indian society inspired by the reading of Gramsci. In the preface Guha states that he is interested in the relationship between the Indian bourgeoisie and the 'people,' a term he uses synonymously with the phrase 'subaltern classes' (Guha, 1982:3). This simplified conceptual framework opens the way to an exploration of the Indian people as opposed to the elite, a history of the subaltern classes constituting the mass of the population. These subaltern groups had roots going back to precolonial times, but were by no means archaic. While India's elite political class had been destroyed by colonialism, the subaltern classes continued to operate as a vigorous part of Indian society. Yet their lives were little known, or only known through a distorted lens. It is for this reason that the Gramscian approach was valuable, with obvious linkages to earlier Italian explorations of subaltern cultures. Thus, while subaltern studies in Italy experienced decline in the 1980s, by a quirk of history they are now experiencing a revival in the writings of Indian scholars.

This early use of Gramsci in Indian social sciences has been carried forward in the sophisticated treatment of what Ananya Roy (2011) calls subaltern urbanism. Roy is especially interested in the epistemic violence of constituting the colonial subject as the Other, in an analysis she draws from readings of Gramsci, and other radical sources. This is a powerful issue owing to voyeuristic media attention which brutally distorts the image of marginal communities in India, refracted through the thin rhetoric of economic neoliberalism. Slum tourism completes the paradigm, where middle-class tourists experience misery and hope at close quarters. Here is a presentation of slum tourism drafted by an author who is clearly enthusiastic about the experience.

The few hours I spend touring Mumbai's teeming Dharavi slum are uncomfortable and upsetting, teetering on voyeuristic. They are also among the most uplifting of my life.

Instead of a neighborhood characterized by misery, I find a bustling and enterprising place, packed with small-scale industries defying their circumstances to flourish amidst the squalor. Rather than pity, I am inspired by man's alchemic ability to thrive when the chips are down.

(Quoted in Roy, 2011:223)

This privileged tourist knows nothing about these poor communities, nor entertains the most remote commitment to their advancement. The tour's perceived excitement hinges on the narrative polarization between hope and despair, mirroring the dominant vision of marginality. The same narrative dynamic is found in the enormously successful film *Slumdog millionaire* (Boyle, 2008), whose director is noted especially for his British comedies, with a producer excelling in light entertainment. It offers decontextualized images of poverty contrasted with the salvific simulacra of entrepreneurial success, in a plot propelling the protagonist from misery to affluence. The protagonist is transformed into the mythic hero of hackneyed portrayals of the rags to riches tale, all geared to the interests of a mass hetero-directed audience. The portrayal dances on the surface of complexity, fueling the moral indifference which perpetuates injustice. The portrayal is diametrically opposed to the fine-grained reading and open plot of Neorealist films, including those of Satyajit Ray, with their authentic concern for the perspective of the marginal poor (Cooper, 2000).

As an antidote to this voyeuristic vision of poverty, Roy proposes subaltern urbanism. Referencing the work of the Indian Subaltern Studies Collective, she claims that subaltern urbanism 'seeks to confer recognition on spaces of poverty and forms of popular agency that often remain invisible and neglected in the archives and annals of urban theory' (Roy, 2011:224). Butola (2019) advances the surprising claim that Roy's use of Gramsci is more epistemic than transformative, yet it is impossible to imagine any transformation separated from a process of recognition. The interaction between knowing and acting is constitutive of Gramsci's project, as we can see in the Roman periphery, where the response to the experience of the self is articulated within the context of changing self-awareness. Roy proposes to disrupt the 'ontological and topological readings of subalternity,' instead seeing marginality in the framework outlined in Gramsci's six points, especially in terms of the autonomy of subaltern groups (Roy, 2011:235). Pasolini believes that a primordial sacred condition somehow unites these communities, but in sociological terms a more probable uniting force is the collective response to the image of the Other promoted by dominant society. Like Pasolini's idea of the sacred, the response is slippery owing to its linkage with the dominant image of marginality.

While Ananya Roy stresses that global cities are not the same as Euro-American cities (Roy, 2009), some of the dynamics she describes in India are not foreign to Rome. Even Butola contrasts the planned cities of the developed world against the informality of Indian megacities (Butola, 2019:8). Yet the situation in Rome is not categorically different. Roy's (2005) description of urban informalism in subcontinental Asia, with its state of ambiguity and exception, is hardly alien to Rome. Even her discussion of Indian development mafias strikes a familiar chord. But notwithstanding similarity, the scale of urban phenomena in Asia is entirely different. One need only consider the six million forceful evictions in the slums of Mumbai, or the dispossession of three million houses in Beijing (Butola, 2019:4). Even when the O zones were created in Rome in the 1970s to define areas of informal growth, the numbers involved were minute by comparison, involving a population of some three or four hundred thousand residents. Nor was Rome, at least in the postwar period, subject to the brazen policy measures evident in these accounts, with the mass destruction of the informal homes of the urban poor. I am thinking of Butola's description of goon squads called in to smash informal homes on the Seoul periphery (Butola, 2019:4). These Asian slums are clearly outside the system of entitlement and recognition, while in Rome the periphery is outside but also inside. In Rome there is commitment to the periphery as a special place, echoing the idea of the sacred we have explored throughout these pages. While interpreted differently by different groups, commitment to place cuts across the political spectrum, and has relevance throughout Rome's urban territory.

A factor driving this commitment is a civic tradition with deep historical roots. But it may also be that Rome's modern growth contains an element of topological stability which has furnished a basis for resistance. Transformations in Rome are often slow – generally too slow by local expectations. Administrative action is hesitant, and in most cases at least puts on a show of attempting to serve the needs of a mixed community. When in an interminable process, after many years of struggles, borgata Gordiani was finally demolished, effort was made to rehouse dislocated residents within the same neighborhood from which they were evicted. Like many instances of demolition and relocation these fall below the level of archival recognition. Former residents of Borghetto Latino still remember the shanties which arose in the 1920s as a response to the fascist gutting out of the historic center. These shacks were only gradually dismantled as new housing became available. The shacks along the Alessandrino Aqueduct followed a similar trajectory, having gained public recognition through the writings of a street priest who established a school there to serve the needs of indigent residents (Sardelli, 1971). Even there it appears that dislocated residents were helped to find new lodgings, often within the same general neighborhood.

Concern for the poor is a commitment expressed in several articles of the Italian constitution. It is part of a Catholic tradition, as well as reflecting the orientation of the country's leftist history. In the postwar period the concern is expressed through the abundance of writings and films concerning marginal communities. This is part of an effort to go beyond the image of the Other generated by the mechanisms of domination. The framework of subalternity has separated but also united interconnections between center and periphery, militating against voyeuristic portrayals.

There is an inevitable effort to colonize the periphery through gentrification and market development, to capture alternative value and put it to the service of the mainstream economy. But there is also resistance to top down urban intervention. The debate triggered by trivializing media portrayals of Indian slums also prompted strong reactions to the presumption of starchitects to transform the slums of the global south according to a vision which ignores entirely the sensitivities of slum residents. In this vision the slum becomes a legal-lethal space suspending the ontological status of its citizens, a plaything in the hands of well-funded developers and their starchitects whom Ananya Roy describes as being delirious with the power of their gaze (Roy, 2011:227). This speculative vision of poverty ignores the political and historical forces that shape the city, treating the urban periphery as a tabula rasa awaiting the creative intervention of the global market (Godlewski, 2010:17).

In the Roman periphery there is still capacity to resist the decontextualized incursion of the global market. One can take the example of the new stadium proposed for AS Roma. The project brings together the greatest architects of the world, funded for the most part by global real estate developers. The debate of whether to move forward with this significant investment in the periphery has gone on for years, and at the moment of this writing has failed to move forward owing to many factors, including the resistance of a significant community segment which has no sympathy for such an intrusive project.[10] These are not romantic self-organizing communities, but communities with sufficient agency and cultural autonomy that they can still impact the shape of the city. Rome has grown in a gradual process balancing planning with informality, fusing together disparate styles and uses in such a way that no element dominates the others. It is no longer the Rome of Pasolini, but it still possesses a memory which is a stimulus to celebrate the life of the city even in simple ways. Alongside major political debates, one finds diffuse micro-activism intertwining place and human agency as complementary actants in an ever-changing assemblage. This is the premise for a performance which is urban life itself: when the performance stops, so the city ceases to exist. Fortunately for Rome, that day still seems far off.

Notes

1 https://www.decrescitafelice.it/ (Accessed June 10, 2020).
2 https://www.beppegrillo.it/passaparola-una-rivoluzione-culturale-per-salvare-lumanita-di-serge-latouche/ (Accessed June 10, 2020).
3 http://www.semidicomunita.it/ (Accessed June 10, 2020).
4 Cooperative Coraggio is an acronym for Cooperativa Romana Agricoltura Giovani (Roman Youth Agriculture Cooperative) (https://www.coop-coraggio.it/ - Accessed May 10, 2020).
5 https://libcom.org/library/pier-paolo-pasolini-interviewed-louis-valentin-1970 (Accessed May 1, 2020).
6 https://napoliteatrofestival.it/spettacolo/partire-domani/ (Accessed May 1, 2020).
7 In 2017 the Italian happy degrowth movement published on its site an account of this poem, calling it 'prophetic.' The poem is linked to other themes which interested Pasolini, including his belief that the left rejected the New Testament because the martyrdom of the underclass would turn it into a new Christ. https://www.decrescitafelice.it/2017/04/profezia-una-poesia-di-pier-paolo-pasolini/ (Accessed May 1, 2020).
8 http://africultures.com/personnes/?no=5527 (Accessed June 20, 2020).
9 https://english.ahram.org.eg/NewsContent/5/32/303044/Arts--Culture/Film/Remembering-Salah-AbuSeif-Egypt%E2%80%99s-greatest-realist.aspx (Accessed June 20, 2020).
10 http://www.salviamoilpaesaggio.roma.it/2016/06/salvare-tor-di-valle-dal-cemento-no-al-progetto-stadio/ (Accessed September 20, 2020).

Bibliography

Benhadj, Rachid (dir.) (2006) *For bread alone*. Rome: AE Media Corporation.

Bernini, Stefania (2015) *Pasolini: the sacred flesh*. Toronto: University of Toronto Press.

Bevilacqua, Piero (2014) *Pasolini. L'insensata modernità*. Milan: Jaca Book.

Bobbio, Norberto (1994) *Destra e sinistra. Ragioni e significati di una distinzione politica*. Rome: Donzelli Editore.

Butola, Balbir Singh (2019) 'Subaltern urbanism,' *The Wiley Blackwell Encyclopedia of Urban and Regional Studies*, 1–10.

Boyle, Danny (dir.) (2008) *Slumdog millionaire*. London: Caledor Films.

Cellamare, Carlo (2018) 'L'azione pubblica tra malgoverno e autoorganizzazione,' in Coppola, Alessandro and Gabriella Punziano (eds.) *Roma in transizione: governo, strategie, metabolismi e quadri di vita di una metropoli*, pp. 359–370. Rome-Milan: Planum Publisher.

Celata, Filippo, Raffaella Coletti, Chary Y Hendrickson and Venere Stefania Sanna (2018) 'La città nella crisi. Pratiche informali di resilienza nella Roma post-politica,' in Coppola, Alessandro and Gabriella Punziano (eds.) *Roma in transizione: governo, strategie, metabolismi e quadri di vita di una metropoli*, pp. 373–386. Rome-Milan: Planum Publisher.

Cherki, Eddy and Michel Wieviorka (2007) 'Autoreduction movements in Turin,' in Lotringer, Sylvère and Christian Marazzi (eds) *Autonomia: post-political writings*, pp. 72–79. Los Angeles: Semiotext(e).

Cooper, Darius (2000) *The cinema of Satyajit Ray: between tradition and modernity*. Cambridge, UK: Cambridge University Press.

Coppola, Alessandro (2012) *Apocalypse town: cronache dalla fine della civiltà urbana*. Bari: Laterza.

Cuffaro, Lucia (2014) *Fatto in casa. Smetto di comprare tutto ciò che so fare.* Bologna: Arianna Editrice.

Della Porta, Donatella (ed.) (2004) *Comitati di cittadini e democrazia urbana.* Soveria Mannelli: Rubbettino Editore.

Gentili, Dario and Elettra Stimilli (2015) *Differenze italiane. Politica e filosofia: mappe e sconfinamenti.* Rome: Labirinti.

Gobetti, Sandro and Luca Santini (2018) *Reddito di base, tutto il mondo ne parla. Esperienze, proposte e sperimentazioni.* Florence: GoWare.

Godlewski, Joseph (2010) 'Alien and distant: Rem Koolhaas on film in Lagos, Nigeria,' *Traditional Dwellings and Settlements Review*, 21:2:7–20.

Gray, Neil (2018) 'Beyond the right to the city: territorial autogestion and the take over the city movement in 1970s Italy,' *Antipode*, 50:2:319–339.

Gramsci, Antonio (1982 [1971]) *Selections from the prison notebooks.* Edited and translated by Quintin Hoare and Geoffrey Nowell Smith. London: Lawrence and Wishart.

Guha, Ranajit (1982) *Subaltern studies. Writings on south Asian history and society 1.* Oxford: Oxford University Press.

Herzfeld, Michael (2009) *Evicted from eternity: the restructuring of modern Rome.* Chicago: University of Chicago Press.

Latour, Bruno (2007) *Reassembling the social: an introduction to actor-network-theory.* Oxford: Oxford University Press.

Latouche, Serge (2015) *Farewell to growth.* Translated by David Macey. Cambridge, UK: Polity Press.

Lefebvre, Henri (1996) 'Right to the city,' in Kofman, Eleonore and Elizabeth Lebas (eds.) *Writings on cities*, pp. 61–181. Oxford: Blackwell.

Lefebvre, Henri (2009) 'Theoretical problems of autogestion,' in Brenner, Neil and Stuart Elden (eds.) *State, space, world: selected essays by Henri Lefebvre*, pp. 138–152. Minneapolis: University of Minnesota Press.

Lotta Continua (1973) 'Take over the city: community struggles in Italy,' Translated and edited by Ernest Dowson. *Radical America*, 7:2:79–112.

Macri, Tito (2016) 'Su alcuni sviluppi del concetto di "società politica". Uno sguardo ai subaltern studies,' *Sociologia. Rivista Quadrimestrale di Scienze Storiche e Sociali*, 2:123–186.

McFarlane, Colin (2011) *Learning the city. Knowledge and translocal assemblage.* Sussex: Wiley-Blackwell.

Mosco, Valerio Paolo (2018) 'Roma moderna. Una proposta per la Capitale,' *Artribune*, 4 March.

Nicolini, Renato (2011) *Estate Romana: un effimero lungo nove anni.* Reggio Calabria: Città del Sole Edizioni.

Pasolini, Pier Paolo (dir.) (1968) *Teorema*. Rome: Aetos Produzioni Cinematografiche.

Pasolini, Pier Paolo (1976 [1964]) *Poesia in forma di rosa.* Milan: Garzanti.

Pasolini, Pier Paolo (1989 [1983]) *Pier Paolo Pasolini: il sogno del centauro.* Edited by Jean Duflot. Rome: Editori Riuniti.

Pasolini, Pier Paolo (1990 [1974]) *L'odore dell'India.* Milan: Garzanti.

Pasolini, Pier Paolo (2005) *Heretical empiricism.* Translated by Ben Lawton and Louise K. Barnett. Washington, D.C.: New Academia Publishing.

Pasolini, Pier Paolo (2018 [1975]) *Scritti corsari.* Milan: Garzanti.

Pileri, Paolo (2020) 'Cara Task force, date prova di ripresa nel cambiamento e non solo di ripresa,' *Altreconomia*, 18 April.

Pontale, Dario (2019) *La Roma di Pasolini. Dizionario urbano.* Rome: Nova Delphi.

Roy, Ananya (2005) 'Urban informality: toward an epistemology of planning,' *Journal of the American Planning Association*, 71:2:147–158.

Roy, Ananya (2009) 'The 21st-century metropolis: new geographies of theory,' *Regional Studies*, 43:6:819–830.

Roy, Ananya (2011) 'Slumdog cities: rethinking subaltern urbanism,' *International Journal of Urban and Regional Research*, 35:2:223–238.

Ruberto, Laura E. and Kristi M. Wilson (2017) 'Italian neorealism: quotidian storytelling and transnational horizons,' in Burke, Frank (ed.) *A companion to Italian cinema*, pp. 139–156. Oxford: Blackwell.

Said, Edward (1994) *Culture and imperialism*. New York: Vintage Books.

Sapelli, Giulio (2015) *Modernizzazione senza sviluppo. Il capitalismo secondo Pasolini*. Florence: GoWare.

Sardelli, Roberto (1971) *Scuola 725: non tacere*. Florence: Libreria Editrice Fiorentina.

Smith, Gregory and Gilda Berruti (2019) 'Social agriculture, antimafia and beyond: toward a value chain analysis of Italian food,' *Anthropology of Food*, 13 October.

Trento, Giovanna (2010) *Pasolini e l'Africa, l'Africa di Pasolini: panmeridionalismo e rappresentazioni dell'Africa postcoloniale*. Udine: Mimesis.

Ventura, Raffaele Alberto (2017) *Teoria della classe disagiata*. Rome: Minimumfax.

Virno, Paolo and Michael Hardt (eds.) (1996) *Radical thought in Italy: a potential politics*. Drogheda: Choice Publishing.

Vittoria, Albertina (1990) 'La commissione culturale del PCI dal 1948 al 1956,' *Studi Storici*, 31:1:135–170.

Index

Printed in the United States
by Baker & Taylor Publisher Services

Printed in the United States
by Baker & Taylor Publisher Services